SR Supplements / 17

SR SUPPLEMENTS

Volume 17

Christ and Modernity
Christian Self-Understanding in a Technological Age

David J. Hawkin

Published for the Canadian Corporation for Studies in Religion/Corporation Canadienne des Sciences Religieuses
by Wilfrid Laurier University Press

1985

Canadian Cataloguing in Publication Data

Hawkin, David J.
Christ and modernity

(SR supplements ; 17)
Bibliography: p.
ISBN 0-88920-193-5.

1. Christianity – 20th century. 2. Technology – Religious aspects – Christianity. I. Title. II. Series.

BR121.2.H39 1985 261.5′6 C85-099996-0

85 86 87 88 4 3 2 1

Cover design by Michael Baldwin, MSIAD

Order from:
Wilfrid Laurier University Press
Wilfrid Laurier University
Waterloo, Ontario, Canada N2L 3C5

Printed in the United States of America

To my late father, JOHN WILLIAM,

and to my mother, JESSIE

Vivendo, immo moriendo et damnando fit Christianus,
non intelligendo, legendo, aut speculando

Table of Contents

Acknowledgements vii
Abbreviations viii
Introduction ix

PART I: THE PHENOMENON OF CHRISTIANITY

Chapter One
The Origin: Jesus 3
1. Hermann Samuel Reimarus: Jesus the Revolutionary 4
2. David Friedrich Strauss: Jesus in Mythical Garb 8
3. Albert Schweitzer: Jesus the Eschatological Figure 13
4. Rudolf Bultmann: Jesus the Proclaimer of the Time of Decision 20
5. The Mysterious Christ 31

Chapter Two
The Development: Belief and Practice in the Early Church 34
1. Orthodoxy, Heresy, and Development: The Semantic Problem 38
2. A Point of Departure: The *Lex Orandi* 40
3. "Development" Revisited 43
4. The Distinctively Christian in the New Testament 46
5. The *Lex Orandi* and the Early Christian Style of Life 49

Conclusion to Part I
Understanding the Phenomenon of Christianity 52

PART II: THE PHENOMENON OF MODERNITY

Chapter Three
The Origin: Changing Horizons 59
1. The Horizon of Suspicion 61
2. The Horizon of Modernity 73

Chapter Four
The Development: Towards a Technological Future
1. Some Theological Perspectives on Modernity and Christianity 80
2. A Dissenting View: Technique and the Eclipse of Human Autonomy 85

Conclusion to Part II
Understanding the Phenomenon of Modernity 94

PART III: CHRIST AND MODERNITY

Chapter Five
Some Basic Issues
1. Continuity and Discontinuity in Christianity and Modernity 99
2. The Autonomy of Humanity 103
3. Technology and Christianity: Some Biblical Reflections 106

Conclusion
Christian Self-Understanding in a Technological Age 117

NOTES

Notes to Chapter One 123
Notes to Chapter Two 132
Notes to Conclusion to Part I 142
Notes to Chapter Three 145
Notes to Chapter Four 150
Notes to Conclusion to Part II 156
Notes to Chapter Five 157
Notes to Conclusion 163

Bibliography 164
Index 178

Acknowledgements

Throughout the writing of this book I have received help from many quarters. I particularly wish to thank Professors C. K. Barrett and B. F. Meyer for their gracious assistance and encouragement, and the Revd. B. A. Mastin, who offered many valuable criticisms and suggestions. To my good friends Kathleen Clarkson and Dr. Katharine Temple I owe a special debt of gratitude for their help and advice in writing the sections on Marx and Ellul. I also wish to thank Professors P. C. Craigie and S. G. Wilson for their help in the initial stages of the work.

The editor of *Eglise et Théologie* has kindly granted permission to use extracts from my article "A Reflective Look at the Recent Debate on Orthodoxy and Heresy in Early Christianity," which originally appeared in *Eglise et Théologie* 7 (1976), 367-378. These extracts appear in chapter two. Material from pp. 100-102 has appeared in *Churchman* 99 (1985), pp. 51-56.

To Canada Council I wish to express my gratitude for a Leave Fellowship (no. 451-80-2211), which enabled me to take a sabbatical to work on this book.

Finally, I am most indebted to my wife Eileen not only for her many useful criticisms and suggestions in the course of writing of this book, but also for her invaluable help in preparing the final copy.

This book has been published with the help of a grant from the Canadian Federation for the Humanities, using finds provided by the Social Sciences and Humanities Research Council of Canada.

Abbreviations

BJRL	*Bulletin of the John Rylands Library*
BZ	*Biblische Zeitschrift*
BZAW	Beihefte zur Zeitschrift fuer die Alttestamentliche Wissenschaft
BZNW	Beihefte zur Zeitschrift fuer die Neutestamentliche Wissenschaft
CBQ	*Catholic Biblical Quarterly*
ExTim	*Expository Times*
ET:	English Translation
EvTh	*Evangelische Theologie*
EvQ	*Evangelical Quarterly*
FRLANT	Forschung zur Religion und Literatur des Alten und Neuen Testaments
HarvThR	*Harvard Theological Review*
ICC	International Critical Commentary
JBL	*Journal of Biblical Literature*
JournRel	*Journal of Religion*
JTS	*Journal of Theological Studies*
NovT	*Novum Testamentum*
NTS	*New Testament Studies*
RB	*Revue Biblique*
SBT	Studies in Biblical Theology
SJTh	*Scottish Journal of Theology*
StTh	*Studia Theologica*
TDNT	*Theological Dictionary of the New Testament*
TS	*Theological Studies*
TZ	*Theologische Zeitschrift*
ZNW	*Zeitschrift fuer die Neutestamentliche Wissenschaft*
ZTK	*Zeitschrift fuer Theologie und Kirche*

Introduction

In 1927 Martin Heidegger published his great work *Sein und Zeit* in which he thematized technology as the hallmark of the modern age. Since the publication of his book no one has seriously disputed that technology has shaped the contours of modernity. What is disputed is whether the power of technology is salutary. While it is still the prevalent view that the technological enterprise is an endeavour which benefits humanity, there are those who argue that it will ultimately dehumanize us.

For Christians this is a perplexing dispute, as the particular attitude they should take is not at all clear. Some claim that the technological process is inimical to Christianity, while others affirm that it is derived from, and sustained by, an authentic Christian vision of the world. These conflicting views are rooted in differing presuppositions concerning the nature of both Christianity and modernity. In Part I we will deal with the question of the nature of Christianity by examining its origin and development. In Part II we will look at some of the roots of modernity and try to delineate its most important characteristics. Part III will then deal with some of the basic questions which have arisen from our examination of Christianity and modernity. In particular we will attempt to describe those characteristics of modernity which may be accommodated to Christianity and those which may not. A major focus of our discussion will be whether the Christian understanding of human autonomy is compatible with that which underlies modernity. The conclusion will address the question of how, in the light of our discussion, Christians should understand their faith in our technological age.

PART I

THE PHENOMENON OF CHRISTIANITY

Chapter One

The Origin: Jesus

Christians are so called because they are followers of Jesus Christ, and any discussion of the nature of Christianity must accordingly begin with this fact. Having stated this principle, however, we immediately run into problems. What does it mean to "follow" Jesus Christ? Does following Jesus simply mean adhering to the principles of his teaching? Does his teaching have any relevance for humanity today? Indeed, do we know exactly what Jesus taught? What is implied by the confession of Jesus Christ as "Lord" (Phil 2:11)? Does it imply that the Christian has allegiance to more than an historical person? If so, just how important is the historical Jesus for the Christian?

These and related questions have been the key questions of New Testament scholarship for over two hundred years. The two most basic issues revolve around the quest of the historical Jesus and the relation of the historical Jesus to the Christ of faith. The quest of the historical Jesus has been at the forefront of New Testament scholarship ever since the publication in 1778 of Hermann Samuel Reimarus's *Vom Zweck Jesu and seiner Juenger.*[1] Reimarus challenged the assumption that the Gospels give us an accurate account of the historical Jesus. The publication in 1835-36 of David Friedrich Strauss's *Leben Jesu*[2] raised the further question of whether the historical Jesus was important for the Christian faith. Strauss challenged the assumption that the Gospels intended history. For Strauss they were not historical; rather they were mythical. In Strauss's view this did not mean that Christianity was discredited, for its true nature was not dependent upon the past particulars of history. Reimarus and Strauss are two pivotal figures in the history of New Testament scholarship, for they posed questions which no one had systematically dealt with before. Albert Schweitzer's *Von Reimarus zu Wrede: eine Geschichte der Leben-Jesu-Forschung*[3] (1906) was the first work to recognize their true significance. Schweitzer himself greatly contributed to the debate by arguing that Jesus was a figure of his own time and was "a stranger" to ours. Schweitzer's thesis was that Jesus was an apocalyptic visionary whose outlook and actions had become foreign to us. The greatest New Testament scholar of the modern age, Rudolf Bultmann, attempted to deal with the questions raised by Reimarus, Strauss, and Schweitzer by retranslating the New Testament message into existential categories. It behoves us, therefore, to examine these four New Testament scholars in greater detail, for we cannot begin to discuss the nature of Christianity unless we understand the questions they raised and the answers they gave.

1. Hermann Samuel Reimarus: Jesus the Revolutionary

Between 1774 and 1778 there appeared in *Beitraege zur Geschichte und Literatur* six extracts from a manuscript entitled *Apologie oder Schutzschrift fuer vernuenftigen Verehrer Gottes*. The editor of the journal was Lessing, and he had found the manuscript in Wolfenbuettel library. Lessing did not know who the author of the extracts was, and so he entitled the extracts "Fragments of an Unknown Author." (The author was later established as Hermann Samuel Reimarus, a teacher in Hamburg who had died in 1768.) The final extract was called "On the Aims of Jesus and his Disciples." Since the publication of that extract in 1778, New Testament scholarship has never been quite the same. It caused a sensation at the time -- because of it many "serious and thoughtful" young men abandoned their plans to become clergymen.[4] It continues to exercise influence today, although often in an oblique way, for although Reimarus's basic thesis continues to be restated and dealt with anew,[5] it is sometimes not realized that Reimarus was its originator! The thesis of Reimarus's work, stated in very bald and simple terms, was this: Jesus was a revolutionary who failed, and his disciples salvaged what they could from the disaster by giving out a spiritual interpretation of his life. So stated, Reimarus's thesis loses much of its impact, for the subtlety and ingenuity of Reimarus can only be appreciated by reading him firsthand and grasping his historical context.

It is difficult to evaluate truly the significance of Reimarus's work, and some -- like Schweitzer -- are very laudatory,[6] while others -- like Kuemmel[7] -- downplay his originality and importance. Now it is certainly true that much of what Reimarus said was not new. Reimarus is often credited, for instance, with being the first to introduce the distinction between the historical Jesus and the Christ of faith,[8] but this distinction was actually first noted by Chubb.[9] Schweitzer claimed that Reimarus was the first to point to the importance of eschatology for Jesus, but again this had been noted previously by Semler.[10] Yet Reimarus did make an extremely valuable contribution to New Testament scholarship, for his work crystallized in a remarkable way many of the key issues which have subsequently dominated the discipline.

In the first of six extracts, Reimarus states that he does not want to disturb people. His book was to be kept secret, except from intelligent friends (this did not apparently include his wife!), and was only to be published when the times were more enlightened. He was writing not for publication but to "fully satisfy myself and my rising doubts. I could not but thoroughly investigate the faith that had raised so many difficulties for me in order to discover whether it could subsist with the rules of truth or not."[11] In the third extract Reimarus establishes the principle that the Bible must not have its meaning imposed on it from without and that it has but one meaning, a simple, unequivocal one. In this, of course, he is simply in accord with the spirit of his post-Reformation age. These two principles are applied in a most rigorous and systematic way in the final and most important extract, "On the Intention of Jesus and His Disciples."

Armed with his commitment to truth and his determination to interpret the Gospels for himself in the cool light of reason, Reimarus sets about his task in a methodical way. His question is: What did Jesus actually intend? It was an historical question posed in a new way. To unravel the mystery of Jesus' intention, Reimarus makes a distinction between apostolic interpretation and the actual Jesus of history.

> Jesus left us nothing in writing; everything we know of his teaching and deeds is contained in the writings of his disciples. Especially where his teaching is concerned, not only the evangelists among his disciples, but the apostles as well undertook to present their master's teaching. However, I find great cause to separate completely what the apostles say in their own writings from that which Jesus himself actually said and taught, for the apostles were themselves teachers and consequently present their own views.[12]

Reimarus accordingly settles on the four Gospels as the documents which will reveal most about Jesus' intention:

> Now since there are four of them and since they all agree on the sum total of Jesus' teaching, the integrity of their reports is not to be doubted. . . . I have sufficient reason to limit myself exclusively to the reports of the four evangelists who offer the proper and true record.[13]

In his examination of the four Gospels, Reimarus finds that Jesus teaches the Kingdom of God, a new morality, the need for repentance, and the coming judgment. This teaching was entirely in accord with Jewish teaching generally; even in his proclamation of a new ethic, Jesus had no intention of introducing a new religion -- he "kept the feasts" and intended to preach only to Jews. There is nothing in the Gospels, claims Reimarus, about the trinity[14] or salvation through Jesus Christ, the Son of God. Jesus may have used the title Son of God but merely in the Old Testament sense of "one beloved by God." The famous verse in Jn 10:30, "I and the Father are one," is simply an expression of mutual love.

The key question, then, becomes: What did Jesus mean by the Kingdom of God? For Jesus it meant the driving out of the Romans and the establishing of the reign of the messiah. There can be no doubt that this is what the disciples expected Jesus to do -- "We trusted that it would be he who would redeem Israel" (Lk 24:21).[15] Nor can there be any doubt that this is how Jesus understood himself; he thought of himself as the messiah who would bring in the Kingdom. Hence the entry into Jerusalem is full of messianic significance. Jesus is publicly claiming his messiahship. But, unfortunately for Jesus, no prominent Jew accepts his claim. Faced with failure, Jesus tries to hide but is betrayed by Judas. He is tried, convicted, and crucified as a messianic pretender. He dies a broken man.

Reimarus devotes a good deal of time and effort to a discussion of the resurrection. He finds the accounts of the resurrection to be contradictory and implausible. The argument that Jesus worked miracles and that the resurrection was the consummate miracle is unconvincing to Reimarus. The miracles prove nothing: Jesus himself warned against miracle workers who were, in fact, false messiahs. The most obvious explanation of the resurrection is that the disciples stole the body.[16]

The inevitable question arises at this point: If Jesus died a broken man, and if there is no resurrection, how does one explain the phenomenon of Christianity? Reimarus's answer is that the disciples created Christianity. After Jesus was crucified, they stole the body, waited fifty days, and then proclaimed a resurrected Christ who would soon return in glory. The motivation of the disciples in doing this was quite simple. They had followed Jesus in the hope of positions of power and prestige when Jesus established his earthly kingdom. They had forsaken their occupations in this hope. They were now faced with ruin, disgrace, and ridicule.

> If they returned to their original occupations and trades, nothing but poverty and disgrace awaited them. Poverty, because they had forsaken all, particularly their nets, ships and other implements; and, besides, they had grown out of

> the habit of working. And disgrace because they had experienced such a tremendous downfall from their high and mighty expectations, and by their adherence to Jesus had become so familiar to all eyes, that everyone would have jeered and pointed at the pretended judges of Israel and intimate friends and ministers of the messiah, who now had again become poor fishermen and perhaps even beggars.[17]

When he was alive, Jesus had eaten well (Mt 11:19) and had been financially supported (Lk 8:1-3). The disciples, by putting out a spiritual interpretation of his life, hoped to regain not only some personal pride and power but also material comfort, which they, in fact, succeeded in doing (Acts 4-5).

Reimarus's work does, of course, have many flaws. In particular, his explanation of the resurrection is too facile, as is his belief that the disciples suppressed the true intention of Jesus. As Schweitzer commented: "It was, of course, a mere makeshift hypothesis to derive the beginnings of Christianity from a mere imposture."[18]

But Reimarus's great contribution was that he posed new questions and dealt with them in a thoroughly systematic way. He posed the question of the origin of Christianity and differentiated it from the question of the intention of Jesus. After Reimarus, it was no longer possible to assume that the historical Jesus and his intention were easily accessible through the Gospels. Schweitzer's appreciative evaluation of Reimarus gives us an idea of his importance: "His work is perhaps the most splendid achievement in the whole course of the historical investigation of the life of Jesus."[19] Before Reimarus, Spinoza had asked questions about the authorship of the Gospels, their provenance and purpose, and the veracity of their miracle accounts. But Reimarus's effort was the first thorough historical assessment of the biblical record which focused particularly on the historical Jesus. Subsequent scholars, in the period 1780 to 1830, in order to refute Reimarus's reconstruction of events, concentrated more and more on such questions as the tradition that lay behind the Gospels and the order in which the Gospels were written. Much useful progress was made in these fields -- the work of Griesbach and Herder comes to mind -- but the next major leap forward came with the work of David Friedrich Strauss.

2. David Friedrich Strauss: Jesus in Mythical Garb

"In order to understand Strauss one must love him." So begins Albert Schweitzer's description of the life and fate of Strauss; and while one may not agree that love of Strauss precedes understanding him, one cannot read about him without being strongly stirred. His life has a genuine tragic dimension to it. He wrote a great book, which ruined his academic career. He married the famous singer Agnese Schebest and thereby ruined his chances of a fulfilled married life. He was one of those gifted men of whom capricious fate seemed determined to make an example.

He was born in 1808 into a middle-class family. From 1821-25 he attended the pre-seminary institution of Blaubeuren, where he was a student of F. C. Baur. After a brief period as an assistant pastor, he went to Berlin, with the intention of studying under Schleiermacher and Hegel. Hegel was mortally stricken with cholera shortly after he arrived, and he found himself unable to get on with Schleiermacher. In 1832 he therefore took up a position of assistant lecturer at the theological college.

He was an excellent lecturer. He refused to lecture on anything but philosophy and exhibited in his lectures the influence of Hegel. Although he did not lecture in theology (even though pressured to do so), it was during this period he wrote *Leben Jesu.* The first volume appeared in 1835, and the second volume followed a year later.

Leben Jesu is possibly the greatest book of its type. Certainly Schweitzer thought so and described it as "one of the most perfect things in the whole range of learned literature."[20] In two very large volumes, Strauss -- at that time only twenty-seven -- proceeded to take to task the two major interpretative schools of his day: the rationalists and the supernaturalists. He did it with such attention to detail, such acumen and insight, that the book completely changed the face of inquiry into the life of Jesus. It was a brilliant book. And it was Strauss's ruination.

Upon the publication of the first volume of the book, there was a general uproar. Strauss was fired from his position and subjected to virulent attacks. For the next few years he was engaged in writing a series of rebuttals to his critics, some couched in sarcastic tones.[21] Towards the end of the thirties, however, he grew weary of polemical debate, and after a second edition in which he sought to defend himself against his critics, he published a third edition quite different in tone. In this edition he made a series of concessions to his critics and endeavoured to be irenical.

It seemed, fleetingly, that the proffered olive branch might be accepted. Strauss was offered a position in Zurich in 1839. But a petition was circulated which demanded that the offer be withdrawn. The petition received widespread support, the offer was retracted, and Strauss was pensioned off without ever taking up the position. Strauss's irenical spirit was dissipated, and a fourth edition of the book, published in 1840, reverted to the text of 1835-36. He was never rehabilitated to the academic life. He wrote many works, but none equalled the majesty and power of *Leben Jesu.* What was it about the book which caused such a furor? Wherein lay its greatness?

In *Leben Jesu* Strauss set himself the task of destroying both the natural and the supernaturalist interpretations of the life of Jesus. He used Hegelian dialectic; from the opposition of the supernaturalist and natural interpretations there arises a new one -- the mythical.

> It appeared to the author of the work, the first half of which is herewith submitted to the public, that it was time to submit a new mode of considering the life of Jesus, in place of the antiquated system of supernaturalism and naturalism. . . . The new point of view, which must take the place of the ones indicated above, is the mythical. The theory is not brought to bear on the gospel history for the first time in the present work: it has long been applied to particular parts of that history, and is here only extended to its entire tenor. It is not by any means meant that the whole history of Jesus is to be represented as mythical, but only that every part of it be subjected to a critical examination, to determine whether it have not some admixture of the mythical. The exegesis of the ancient church set out from the double proposition: first, that the Gospels contained a history, and second, that this history was a supernatural one. Rationalism rejected the latter of these presuppositions, but only to cling the more tenaciously to the former, maintaining that these books present unadulterated, though only natural, history. Science cannot rest satisfied with this half-measure: the other presupposition must also be relinquished, and the inquiry must first be made whether in fact, and to what extent, the ground on which we stand in the Gospels is in any way historical.[22]

The mythical interpretation of the Gospels was not new, as Strauss acknowledged, but it had never been applied in a rigorous manner to the Gospels. It was perhaps inevitable that in so doing Strauss would be sorely misunderstood. He did attempt to explain what he meant by myth. Put simply, it was the narrative embodiment of an idea.[23] But for those theologians with a limited understanding of, or who were simply unable to accept, Hegel's distinction between form and idea, Strauss's claim that Christianity should be unaffected by his work must have seemed somewhat disingenuous.

> The author is aware that the essence of the Christian faith is perfectly independent of his criticism. The supernatural birth of Christ, his miracles, his resurrection and ascension, remain eternal truths, whatever doubts may be cast on their reality as historical facts. . . . [I]n the meantime let the calmness and insensibility with which, in the course of it, criticism undertakes apparently dangerous operations, be explained solely by the security of the author's conviction that no injury is threatened to the Christian faith.[24]

In defence of his interpretation of the Gospels as mythical, Strauss points to the work of Gabler and Eichhorn on the Old Testament. The Gospels had previously been shielded from a mythical interpretation because it had been supposed that they had been written by eyewitnesses. But this view was no longer tenable.

> But this alleged ocular testimony, or proximity in point of time of the sacred historians to the events recorded, is a mere assumption, an assumption originating from the titles which the biblical books bear in the Canon. . . . But that little reliance can be placed on the headings of ancient manuscripts, and of sacred records more especially, is evident, and in reference to biblical books has long since been proved. . . . It is an incontrovertible position of modern criticism that the titles of the biblical books represent nothing more that the design of their author or the opinion of Jewish or Christian antiquity respecting their origin.[25]

Strauss further presses home his point by acknowledging those who had courageously found myth in the birth narratives and at the end of the Gospel story. They had been courageous in acknowledging myth to be present here, but they were not courageous enough. Myth was part and parcel of the *whole* Gospel narrative.

> This writer applies the notion of the mythus to the entire history of the life of Jesus; recognizes mythi or mythical embellishments in every portion, and ranges under the category of the mythus not merely the miraculous occurrences during the infancy of Jesus, but also of his public life; not merely miracles operated on Jesus, but those wrought by him.[26]

Strauss argues, as Bultmann was to later, that myth was the mode of perception of the prescientific mind. The Evangelists, when writing of the life of Jesus, would conceive it in mytho-poetic terms. Their intent was not fraudulent.[27]

The final pages of Strauss's introduction detail the canons of historical investigation which determine where the mythical is present. These canons consist of negative and positive criteria and foreshadow in a remarkable way the later efforts of form criticism. Having thus prepared the groundwork, Strauss then turned to the examination of the life of Jesus and demonstrated the superiority of the mythical interpretation over the naturalist and the supernaturalist.

Meticulous in detail, precise in statement, and devastating in conclusion, Strauss's treatment of the Gospel narratives is superb. Unfortunately, his contempt for the interpretations of many of his contemporaries is too ill-concealed -- he has a section called "Sea Stories and Fish Stories" -- and he stands accused of being unduly provocative. Moreover, his treatment is so detailed that often his readers could not see the forest for the trees. It was generally thought, for example, that Strauss had intended to show that it was impossible to say anything about the historical Jesus. But this is not the case. Careful gleaning of what Strauss says does, in fact, yield an (albeit sketchy) outline of the historical Jesus.[28] Strauss never intended to say that Jesus was a mythical figure. As Schweitzer says:

> To assert that Strauss dissolved the life of Jesus into myth is, in fact, an absurdity which, however often it may be repeated by people who have not read his book, or have read it only superficially, does not become any the less absurd by repetition.[29]

Jesus is, however, presented to us in mythical garb.

There are, of course, many criticisms that one can make of Strauss's work, and there has been no shortage of people to point them out. He extends the body of the mythical too far. He does not understand the relation of the Synoptics to each other (in particular, he resisted throughout his life the idea of Marcan priority). He, like Reimarus before him, does not adequately explain the origin of Christianity nor the uniqueness of Jesus. It is difficult to believe that the shadowy Jesus whom Strauss silhouettes for us could launch a faith which captivated so many subsequent generations and cultures.[30] These are all standard criticisms of Strauss and have to be frankly acknowledged as apposite. But the originality of the work far outweighs these piecemeal criticisms. Before Strauss, it had been widely assumed that the Gospels were historical sources for the life of Jesus. Reimarus had assumed this; he had maintained that the intent was historical but that it was a fraudulent intent -- the Gospels did not reveal the true history of Jesus. Strauss shifted the ground of supposition: the Gospels were not historical by intent; they were mythical. They did not belong to the genre of history. Strauss thus raised a whole series of issues which, although sidestepped for a time, had to be dealt with sooner or later: especially the relation of the Jesus of history to the Christ of faith and the historical accessibility of Jesus. New Testament scholarship immediately after Strauss never really grasped the full significance of Strauss's achievement. In fact, during the nineteenth century the most dominant school of New Testament interpretation was the so-called Liberal school. The Liberals virtually ignored the work of Strauss, and in some respects the Liberal lives of Jesus "represent a continuation of interests of the earlier rationalists."[31] The intent of the Liberals was laudable enough: they wished to present Jesus in a form which, while acceptable to the modern world, would also challenge its religious sensibilities. They believed that it was possible to discover the basic outline of Jesus' life and teaching and that the centre of Jesus' message was a timeless ethic: the Fatherhood of God and the brotherhood of humanity. This can be clearly seen in Adolf Harnack's *Das Wesen des Christentums*,[32] a book which was translated into many languages and sold thousands of copies. Harnack summarized the "leading features of Jesus' message" under the following headings: the Kingdom of God and its Coming; God the Father and the Infinite Value of the Human Soul; the Higher Righteousness and the Commandment of Love.

The most influential exponent of Liberal Theology was not, however, Harnack, but Albrecht Ritschl, who conceived of the Kingdom of God "in purely ethical terms." Perrin says the following about Ritschl's position:

> Jesus saw in the Kingdom of God the moral task to be carried out by the human race, . . . it is the organization of humanity through action inspired by love. Christianity itself is therefore both completely spiritual and thoroughly ethical.[33]

Although the complete demise of Liberal Theology did not come until the First World War, its death knell was, in fact, sounded by Albert Schweitzer at the turn of the century. It is to Schweitzer that we now turn.

3. Albert Schweitzer: Jesus the Eschatological Figure

Albert Schweitzer is perhaps the most fascinating scholar of the twentieth century. A man of truly formidable talents and indefatigable energy, he stands like a colossus on the horizon of historical Jesus research. His monumental book *The Quest of the Historical Jesus*,[34] although written shortly after the turn of the century, is still a classic and invaluable reading for anyone wishing to understand the issues involved in attempting the recovery of the historical Jesus.

Much of Schweitzer's attraction lies in the fact that he was more than a New Testament scholar -- much more. Besides writing the dominant book of the age in historical Jesus research,[35] he became a world expert on Bach and organ music, a renowned medical doctor who devoted much of his life to the hospital at Lambaréné, and a recipient of a Nobel peace prize. He is just too big to handle. This is undoubtedly why besides having his adulators he also has his detractors. Truly great men invariably elicit strong emotions of one type or another. To some Schweitzer is a saintly man and to others a despotic tyrant.[36] There is so much of him, you can make of him almost what you will.

For our purposes, however, it is possible to cut him down to manageable size, for our interest in him centres on his historical Jesus research. For those interested in his life, there is Schweitzer's own word on the subject[37] and many biographies.

Schweitzer's *Quest of the Historical Jesus* has to be seen in the context of its time. The book is a summary of the lives of Jesus from Reimarus to Wrede, together with Schweitzer's own reconstruction. The summary of the previous lives of Jesus is a remarkable achievement. Schweitzer's lively style and use of picturesque images and metaphors cannot fail to keep even the most obtuse reader alert to the basic issues. The reader is led into the maze of conflicting reconstructions and interpretations and emerges at the other end thinking how easy and self-evident were all the choices of direction! Schweitzer's own reconstruction of the life of Jesus does not evince the same sense of certainty and of being self-evident, however, and has been the subject of much criticism. To attempt to give a summary of Schweitzer's own summary of previous scholarship would be pretentious indeed. Only by reading it for oneself can one fully appreciate it. Schweitzer's own reconstruction of the life of Jesus is, however, more susceptible of summary and discussion.

Schweitzer saw the nineteenth century as having posed three choices: either historical or supernatural, either the Synoptics or John, either eschatological or non-eschatological.[38] There were two books in particular which Schweitzer thought to be very important and which greatly influenced his own thinking: Johannes Weiss's *Jesus' Proclamation of the Kingdom of God*[39] and William Wrede's *The Messianic Secret in the Gospels: A Contribution toward the Understanding of the Gospel of Mark.*[40]

Schweitzer is full of praise for Weiss, who was the first to tender a "thoroughgoing" eschatological understanding of the teaching of Jesus.

> In passing . . . to Johannes Weiss the reader feels like an explorer who after wanderings through billowy seas of reed-grass at length reaches a wooded tract, and instead of swamp feels firm ground beneath his feet, instead of yielding rushes sees around him the steadfast trees. At last there is an end of "qualifying clause" theology, of the "and yet," the "on the other hand," the "notwithstanding!" The reader had to follow the others step by step, making his way over every foot-bridge and gang-plank which they laid down, following all the meanderings in which they indulged, and must never let go their hands if he wished to come safely through the labyrinth of spiritual and eschatological ideas which they supposed to be found in the thought of Jesus. In Weiss there are none of these devious paths: "Behold the land lies before thee."[41]

In his book Weiss took to task the Liberal understanding of the Kingdom of God. The Kingdom of God in the teaching of Jesus was not a moral organization of humanity which could be brought about by loving one another. Rather it was a transcendental reality, without political overtones, and initiated by God himself *in the future.* Ritschl's understanding of the Kingdom of God is simply at odds with Jesus' understanding of it. Interestingly enough, Weiss thought that Ritschl's understanding of the Kingdom of God, although not congruent with that of Jesus, was still the best for contemporary Christianity! In the preface to the second edition of his book he writes:

> I am still of the opinion that his [Ritschl's] theological system, and especially this central concept [of the Kingdom of God], presents that form of teaching concerning the Christian faith which is most effectively designed to bring our generation nearer to the Christian religion; and, properly understood and rightly used, to awaken and further a sound and strong religious life such as we need today.[42]

This is a very interesting example of the "compartmentalizing" of the academic and the devotional. Weiss's own historical studies did not seem to have any effect upon his own religious sensibilities and perception of Jesus. In this respect his work is inferior to that of Schweitzer, who goes beyond Weiss in applying eschatological categories not only to the teaching of Jesus but also to Jesus' own conception of his life and mission. Moreover, Schweitzer was not oblivious to the question of how an eschatological Jesus -- so foreign to the modern world -- could possibly demand allegiance.[43]

If Weiss represents thoroughgoing eschatology, Wrede represents thoroughgoing scepticism. Wrede took to task those New Testament scholars who had seen in Mark, the earliest Gospel,[44] a reliable historical outline of the life of Jesus.[45] Wrede focuses upon the so-called messianic secret in the Gospels. Wrede's observations have that quality of all original observations -- once made, they seem obvious. There is obviously a messianic secret in the Gospel of Mark (and therefore in the parallels in Matthew and Luke). Jesus silences demons who would make him known (Mk 1:25, 34; 3:11f.); he commands silence after notable miracles (1:44; 5:43; 7:36; 8:26) and after Peter's confession (8:30) and the transfiguration (9:9). He goes on secret journeys (7:24; 9:30) and gives esoteric teaching to his disciples (4:10f.). The data of secrecy are indisputably there -- it is what one makes of such data that is disputable. Wrede claimed that the messianic secret was retrojected into the life of Jesus by the early Church to explain the embarrassing fact that Jesus was not recognized as messiah in his lifetime and, indeed, died ignominiously on a cross. If Jesus were presented as deliberately concealing the fact that he was the messiah, then the lack of recognition by his fellow Jews would be explicable.

Wrede's thesis has had a mixed reception. Some accept it in modified form (Bultmann, for example, thinks that the messianic secret is a redactional device of Mark himself). Others reject its overall conclusions.[46] But it was a provocative thesis and one which could not be ignored. Schweitzer saw it as very significant. In his characteristically pungent style, he says:

> Formerly it was possible to book through-tickets at the supplementary-psychological-knowledge office which enabled those travelling in the interests of Life-of-Jesus construction to use express trains, thus avoiding the inconvenience of having to stop at every little station, change, and run the risk of missing their connexion. This ticket office is now closed. There is a station at the end of each section of the narrative, and the connexions are not guaranteed.[47]

Schweitzer, then, was influenced by the arguments of both Weiss and Wrede. Jesus' life and teaching were determined by eschatology, and the Gospel of Mark was not a straightforward record of the life of Jesus. Schweitzer, however, adapted Wrede's thesis somewhat for his own purposes. There is a messianic secret in the Gospels, but this goes back to Jesus himself. The Gospel of Mark may not be used as unequivocally as previously in reconstructing the life of Jesus, but the historical Jesus was still accessible through Mark and Matthew (especially Mt 10 and 11). "The Life of Jesus cannot be arrived at by following the arrangement of a single Gospel, but only on the basis of the tradition which is preserved more or less faithfully in the earliest pair of Gospels."[48] Schweitzer's most ingenious suggestion -- ingenious in its simplicity -- is that the enigmatic character of the Gospels is due not to the Evangelists but to the character of Jesus himself. Jesus was dominated by "a 'dogmatic idea' which rendered him indifferent to all else."[49] Schweitzer expounded this "dogmatic idea" of Jesus which explains the mysterious course of his life. Jesus' whole teaching was centred around the announcement of the Kingdom of God. Following Weiss, Schweitzer believed Jesus to have understood by this phrase the irruption of the activity of God into human history. Jesus was proclaiming the apocalyptic end of history. He was not the moralizer beloved by the Liberals; the Sermon on the Mount was an "interim ethic," valid only between the time of the proclamation of the Kingdom and its coming. When did Jesus expect the Kingdom of God to arrive? At harvest time. In Mt 9:37f. Jesus spoke of the harvest as plentiful and the labourers as few. He was actually speaking analogically of the harvest in the Palestinian fields and the harvest of the Kingdom of God. The parables reinforce this view; in, for example, the Sower, Jesus was stressing that the seed had been sown. The Kingdom of God was imminent, it must follow "as certainly as harvest follows seed-sowing."[50] It would, in fact, come at the same time as the harvest in the fields.

Schweitzer relied very heavily on Mt 10 and 11, which he regarded as almost wholly historical,[51] for his reconstruction of what happened next. Jesus sent out his disciples to prepare for the advent of the Kingdom at harvest time. He did not expect the disciples back before the Son of Man was made manifest[52] and the Kingdom had come (Mt 10:23). Yet they did return, and the Kingdom had not come.

> The disciples returned to Him; and the appearing of the Son of Man had not taken place. The actual history disavowed the dogmatic history on which the action of Jesus had been based. An event of supernatural history which must take place, and must take place at that particular point of time,

failed to come about. That was for Jesus, who lived wholly in the dogmatic history, the first "historical" occurrence, the central event which closed the former period of His activity and gave the coming period a new character.[53]

Jesus was faced with failure. What he had proclaimed had not happened. Now either he must face ultimate failure and the abandonment of all eschatological hopes, or he must rethink his whole mission. He chose the latter course and embarked upon a most daring and heroic course of action: he himself would usher in the Kingdom by setting in motion the final messianic tribulations.

> [Jesus] in the knowledge that He is the coming Son of Man lays hold of the wheel of the world to set it moving on the last revolution which will bring all ordinary history to a close. It refuses to turn, and He throws Himself upon it. Then it does turn; and crushes Him. . . . The wheel rolls onward, and the mangled body of the one immeasurably great Man, who was strong enough to think of Himself as the spiritual ruler of mankind and to bend history to His purpose, is hanging upon it still. That is His victory and His reign.[54]

Jesus knew himself to be the messiah, but the disciples did not. Jesus had never, in fact, intended to reveal this fact to them; it was "wrung from Him by the pressure of events."[55] How? To explain this, Schweitzer indulged in a little textual rearranging. He placed the transfiguration before the incident at Caesarea Philippi. It is at the transfiguration that the fact that Jesus is the messiah is revealed to the "inner three" disciples; the fact is not revealed by Jesus himself, but intuited by the disciples in a "state of ecstasy."[56] Then, at Caesarea Philippi, the messiahship is revealed by Peter to the rest of the disciples.

Jesus goes to Jerusalem solely for the purpose of dying there. He believes that only his own death will release the messianic tribulations that will bring in the Kingdom. He therefore deliberately antagonizes the Jewish leaders for the purpose of bringing this about.

He is arrested and tried. At the trial the High Priest accuses Jesus of setting himself up as the messiah. Now, according to Schweitzer's thesis, Jesus and the disciples were the only ones who knew the secret of his messiahship.[57] How, then, had the High Priest acquired this knowledge? Through Judas. The question of *why* Judas betrayed Jesus is what has received most attention by scholars; the real question is: *What* was it that Judas betrayed? The answer is that Judas betrayed the messianic secret: he told the High Priest and priests that Jesus thought himself to be the messiah. This explains why, when Pilate offered to release either Jesus or Barabbas, the crowd chose Barabbas. They had been turned against Jesus by the priests who had circulated among them revealing to them the messianic secret. Such a revelation made Jesus appear to be a deluded enthusiast and a blasphemer.

> At midday of the same day -- it was the 14th Nisan, and in the evening the Paschal lamb would be eaten -- Jesus cried aloud and expired. He had chosen to remain fully conscious to the last.[58]

This, then, is Schweitzer's own reconstruction of the life of Jesus. Schweitzer's book was both the "memorial" to the Liberal lives of Jesus and their "funeral oration."[59] What Schweitzer claimed to have done was to return Jesus "to His own time."

> Whatever the ultimate solution may be, the historical Jesus of whom the criticism of the future . . . will draw the portrait can never render modern theology the services which it claimed from its own half-historical, half-modern, Jesus. . . . He will not be a Jesus Christ to whom the religion of the present can ascribe, according to its long-cherished custom, its own thoughts and ideas, as it did with the Jesus of its own making. Nor will he be a figure which can be made by a popular historical treatment so sympathetic and universally intelligible to the multitude. The historical Jesus will be to our time a stranger and an enigma. . . . He passes by our time and returns to His own.[60]

Many criticisms of Schweitzer's thesis have been made. Some of these utilize discoveries and insights which came after Schweitzer and therefore have been made with the benefit of fuller knowledge which progress in scholarship brings. Developments in form criticism and redaction criticism, for example, have given us a more differentiated understanding of the genesis and relation of the Gospel accounts. Redaction criticism has shown that the four Gospels represent four different interpretations of Jesus, each with its own particular viewpoint.[61] Form criticism has shown that Mt 10 is probably an idealized account of a Palestinian missionary venture about A.D. 70, in which the missionary charge is given validity by being put into the mouth of Jesus. The transfiguration, too, is probably a creation of the early Church.[62]

Schweitzer's thesis, then, viewed in the light of later developments in redaction and form criticism, does have serious weaknesses, for the thesis does depend heavily on the historicity of Mt 10 and 11. Schweitzer's work, in fact, exhibits the same vice as that of the Liberals, whom he criticized so well: he made the evidence fit *a priori* notions about Jesus and his life. The origin of Schweitzer's theory about the apocalyptic nature of Jesus' life and mission can be traced back to the very beginning of his theological studies. In 1894, when reading Mt 10 and 11, he received sudden illumination. The insight which came to him at that moment is related thus by Schweitzer himself:

> Matthew 10 contains the account of the Sending out of the Twelve. In the address with which he sent them out Jesus promised them that they would undergo persecution. This however did not happen. He had proclaimed to them also that the Son of Man would come before they had completed their journey through the cities of Israel, which could only mean that with His coming the supernatural Messianic Kingdom would be established. He therefore did not expect them to return. How came it that Jesus proclaimed things to the disciples which did not in fact come to pass? The explanation of Holtzmann, that this address was not from the historical Jesus but was a composition made up of words of Jesus brought together after his death, was for me unsatisfactory. Later, words would not have been put into his mouth which were unfulfilled in the course of the narrative. The text itself forced me to the conclusion that Jesus had in fact expected persecution for the disciples and the coming of the Son of Man, but in the event had been proved wrong. But how did He come to have such an expectation, and what consequences did it have for Him that events proved to be other than He expected?[63]

This experience had for Schweitzer almost "the force of a revelation," and all his subsequent thought was dominated by it.[64] As Perrin says, this is both its strength and weakness:

> Because it came to him with such force he was able to present it forcefully to others, and therein lies its strength. Because it came to him at the very beginning of his studies his later studies of the New Testament and apocalyptic literature were dominated by it, and therein lies its weakness.[65]

Despite these evident criticisms, Schweitzer's achievement was a great one. His presentation of Jesus as "a stranger to our time," the "One Unknown," is so forceful that it opened up the "hermeneutical gulf" between Jesus and ourselves for all to see.[66] Those who still insist that it is only a minor crack and not a yawning chasm attempt the leap across it at their own peril. We might disagree with the details of Schweitzer's reconstruction, we might reject totally the thesis that Jesus was an apocalyptic visionary,[67] but we have to accept that Jesus cannot be appropriated as a man of our time. As Kueng puts it:

> In the light of today's perspectives we have to try to say that what is involved in this immediate [eschatological] expectation is not so much an error as a time conditioned, time-bound world view which Jesus shared with his contemporaries. It cannot be artificially awakened. Nor indeed should the attempt be made to revive it from our very different horizon of experience, although there is always a temptation to do so particularly in what are known as "apocalyptic times."[68]

How was the "hermeneutical gulf" to be bridged? Was it possible to do so without a sacrifice of the intellect? It was this problem which especially dominated the thought of Rudolf Bultmann, to whom we now turn.

4. *Rudolf Bultmann: Jesus the Proclaimer of the Time of Decision*

Anyone who has engaged in serious study of the New Testament is familiar with the work of Rudolf Bultmann. He was undoubtedly the most famous New Testament scholar of the age. His fame was attributable not only to the vast volume of his erudite writings but also to his unique attempt to grapple with the meaning of the New Testament and translate its message into categories meaningful to the modern world. He stood in the tradition of Reimarus, Strauss, Harnack, and Schweitzer; his work incorporates the insights of all of these men, and at the same time he transcends them. Bultmann devoted his life and his considerable talents to trying to bridge the gulf between the modern world and the New Testament. His influence has been such that no one can understand the development of New Testament scholarship without understanding his work.

His life was not as tragic as that of Strauss, nor was it as dazzlingly diversified as that of Schweitzer. This is not to say that his life was untouched by human tragedy. He lost one brother in the First World War and another in the Nazi concentration camps. He was well acquainted with the tragedy and violence of the modern era. Nor would we wish to imply that he was not as talented as Schweitzer. Bultmann was a very gifted man, with a profound understanding of the classics and philosophy. He was also a deeply religious man, who took his preaching seriously. His lucidly simple sermons exhibit a genuine concern to take the Christian message to the ordinary person.

Bultmann was born in 1884, the son of a clergyman. He was a student at Marburg, Tuebingen, and Berlin. In his student days he was influenced by such renowned teachers as Gunkel, Harnack, Juelicher, Weiss, and Hermann. He briefly held academic posts a Marburg, Breslau, and Giessen. In 1921 he was appointed professor of New Testament at Marburg and remained there until his retirement in 1951. He continued to be active in his retirement, however, and besides continuing to write, he gave the Shaffer lectures at Yale in 1951 and the Gifford lectures at Edinburgh in 1955. His death in 1976 was a great loss, for (if we may adapt what W. H. Auden said about Sigmund Freud) he was no more a man than a climate of opinion. His great contribution lies in his consistent and coherent attempt to relate the world of the New Testament to the modern world. He did not shun the challenge of thoroughgoing eschatology but met it head-on. In order to grasp how Bultmann did this, it is necessary to proceed with a few preliminary remarks about form criticism.

At the turn of the century source criticism of the Gospels had generally established the "two-document" hypothesis -- the earliest sources for the life of Jesus that we have are Mark and "Q."[69] Mark, being the earliest Gospel,[70] was generally thought to give an unadulterated and accurate account of the life of Jesus.[71] But the work of Wrede and K. L. Schmidt[72] cast considerable doubt on this view. Schmidt concluded his learned study with the following: ". . . there is therefore no life of Jesus in the sense of an evolving biography, no chronological sketch of the story of Jesus, but only single stories, pericopae, which are put into a framework." The focus of attention then changed to the pericopae themselves -- what was their history, how did they evolve? Was it possible, perhaps, to classify these units according to form or genre and reconstruct the history of the individual pericopae? It was to this task that both Bultmann and Martin Dibelius applied themselves.

Dibelius, in *Die Formgeschichte des Evangeliums*,[73] broke down the units of material comprising the Gospel tradition into various categories: paradigms, novelle (tales), legends, myth, and sayings. It is not necessary for our purposes to examine these categories of Dibelius any further, for our primary concern is with Bultmann, and his classifications are more detailed and complete. In his *Die Geschichte der synoptischen Tradition*,[74] Bultmann divided the Gospel material into two main categories: discourses and narratives. He further broke down the discourses into the following classifications: apophthegms, dominical sayings, parables, and "I" sayings. The narratives he divided into miracle stories, historical narratives, and legends.

The apophthegms are brief anecdotes which centre upon a saying of Jesus; the sole purpose of the narrative portion is to provide a framework for the saying. They are classified as follows: controversy dialogues (e.g., Mk 3:1-6); biographical apophthegms (e.g., Lk 9:57-62); and scholastic dialogues (e.g., Mk 12:28-34). Bultmann concluded that all three types of apophthegms are "ideal" constructions of the Church -- that is, they are not historical. As for the dominical sayings (isolated sayings not accompanying a narrative), they comprise proverbs (e.g., Mk 6:34, Lk 10:7, Mt 22:14); apocalyptic and prophetic sayings (e.g., Mk 1:15, Lk 10:23f., Lk 6:20f.); and sayings concerned with the law and regulations for the community (e.g., Mk 7:15, Mk 3:4, Mt 18:15-17). The proverbs are the "least" likely to be authentic words of Jesus; among the apocalyptic and prophetic sayings there are some authentic words of Jesus; sayings concerned with law and the regulation of the community often do go back to Jesus himself. With regard to parables, while Jesus did use parables, the early Church often transformed them, added explanations, and augmented them with parables from the Jewish tradition. "I" sayings represent the predominantly retrospective viewpoint of the Church.

The narratives Bultmann broke down into miracle stories (e.g., Mk 4:35-41) and historical stories and legends (e.g., Mk 1:12-13). Legends are to be distinguished from miracle stories in that they are edifying stories which may be based on actual happenings.

When confronted with this plethora of classifications and sub-classifications, it is easy to lose sight of the forest because the view is obstructed by trees. Anglo-Saxon scholars, in particular, have tended to be unimpressed with form criticism. T.W. Manson, for example, described it as "interesting but not epoch making."[75] But the implications of form criticism are very far-reaching. Just as the horse trader peers into the mouth of the horse and determines its age by its teeth, so the form critic examines the form of the individual units of the Gospel tradition to determine their ages. How does the classification procedure outlined above help us to do this? Let us take a few examples. Healing miracles follow a set pattern: the nature and extent of the disease is described, then the contact of the healer with the one wishing to be healed, the healing, and then the general astonishment of those witnessing the event. Where there are additions to this pattern, for example, in the giving of the name to the one healed, such additions must be considered secondary.[76] To take another example, in the apophthegms there is an integral connection between the scene and the saying. If the connection appears artificial, then the material is secondary. In other words, these classifications help to establish the pattern of transmission, and where we can spot deviations from this pattern we should become wary of the authenticity of the material.

We can often trace the origin of the Gospel material back to the life of the early Church (the *Sitz im Leben der Kirche*). The form critic is saying that the material used by the Gospel writers was often circulating in the early Church in a fixed oral form before they wrote it down. Some of this material was, in fact, created by the early Church, while other material stemmed from the life of Jesus but was modified to meet the Church's needs. In other words, many of the traditions utilized by the Evangelists reflect the situation of the Church rather than the historical situation in the life of Jesus.

A simple example of this reworking of tradition is found in Mk 12:2-12. In this incident Jesus is asked about divorce. In verses 5-9 Jesus forbids a man to divorce his wife. In Palestine at the time of Jesus a man could divorce his wife, whereas a woman could not divorce her husband. Hence, Jesus does not need specifically to forbid divorce for a woman because he is addressing an audience where this is a legal impossibility. In Mk 10:10-12 the disciples ask Jesus privately to explain his answer. Jesus in his answer to the disciples forbids divorce for either a man or a woman. It seems odd that Jesus would repeat his previously clear answer privately to his disciples. The puzzle is solved when it is realized that Mark is probably writing his gospel in Rome, and in Roman law a woman could divorce a man. It is thought that Mark added verses 10-12 so that Jesus' words applied specifically to his community (the Roman Church).

The implications of form criticism for historical Jesus research are thus far-reaching. Traditions about Jesus circulated orally in the early Church and were modified, adapted, and created by the Church to meet its own specific needs. The early Church transmitted not a chronological, connected account of the life of Jesus, but rather single (unconnected) sayings and narratives (with the possible exception of the Passion narrative). Hence, Bultmann was severely sceptical about attempts to reconstruct the life of Jesus. (Dibelius was a little more positive than Bultmann on this point.) This is not to say that Bultmann thought nothing could be said about the historical Jesus. He says:

> With a bit of caution we can say the following concerning Jesus' activity: Characteristic for him are exorcisms, the breech [sic] of the Sabbath commandment, the abandonment of ritual purifications, polemic against Jewish legalism, fellowship with outcasts such as publicans and harlots, sympathy for women and children; it can also be seen that Jesus was not an ascetic like John the Baptist, but gladly ate and drank

> a glass of wine. Perhaps we may add that he called disciples and assembled about himself a small company of followers -- men and women.[77]

But what we can say about the historical Jesus is very little, and certainly we are unable to reconstruct the intention of Jesus when, for example, he went to Jerusalem; the Gospels "furnish us with no biographical data on the basis of which one can decide what was in Jesus' mind when he went to his death."[78]

Bultmann's form critical work and its conclusions met with criticism, particularly from Anglo-Saxon and Scandinavian scholars.[79] But Bultmann continued to insist that little could -- or indeed *should* -- be said about the historical Jesus. Bultmann rejected the search for the historical Jesus because of his conviction that such a quest contravened the commitment to justification by faith. The object of Christian faith is not the historical Jesus, but the Christ. To emphasize the former leads to a kind of "Jesusology" or hero-worship.

In his justly famous little book *Jesus and the Word*,[80] Bultmann lays down in a few brief pages of introduction his "viewpoint and method" when dealing with the life and teaching of Jesus. It is instructive to look at these few pages.

In the first place, says Bultmann, "our relationship with history is wholly different from our relationship with nature." We cannot "view" history as an impartial observer, because history is not a "museum of antiquities." We must encounter history. History will only speak to people who come to it seeking answers to questions which "agitate" them.[81] Now Bultmann concedes that there are historians who consider themselves "objective." Such historians may well cull many facts *out* of history, but they discover nothing *about* history. Moreover, the so-called objective view of history is often reductionist. An example of such an approach would be the attempt to make Jesus "psychologically comprehensible."

> Now this expression implies that such a writer has at his disposal complete knowledge of the psychological possibilities of life. He is therefore concerned with reducing every component of the event or the personality to such possibilities. For that is what making anything 'comprehensible' means: the reduction of it to what our previous knowledge includes.[82]

Bultmann therefore makes no attempt to make Jesus psychologically explicable. In fact -- and this is the vital point -- Bultmann is not dealing with the past particulars of the history of Jesus at all: "When we encounter the words of Jesus in history . . . *they* meet *us* with the question of how we are to interpret our own existence."[83] In another publication he puts it more pointedly: "I am deliberately renouncing any form of encounter with the Christ after the flesh, in order to encounter the Christ proclaimed in the kerygma, which confronts me in my historic situation."[84] That there was an historical Jesus is of consequence for the Christian faith, but the Christian should not go beyond the "that" of the historical Jesus; to do so is to seek "the 'Christ after the flesh,' who is no longer. It is not the historical Jesus, but Jesus Christ, the Christ, preached, who is the Lord."[85]

It is at this point that we reach the heart of Bultmann's interpretation of Jesus: Jesus' life and work are translated into existential categories. Bultmann's debt to the existentialism of Heidegger is quite evident and openly acknowledged by Bultmann himself.[86] To appreciate Bultmann is, therefore, to appreciate existentialism. Existentialism arose, of course, in reaction to the idea that there are eternal essences. The existentialist begins with our "being-in-the-world," not with some abstract concept of our nature. We make ourselves by our own decisions and actions: such self-determination is our true freedom. This freedom, of course, brings with it responsibility; but to evade it is to languish in inauthentic existence.

Our existence in the world is characterized by feeling; the most important feelings are modes of disclosure. Such a feeling is what Heidegger calls *Angst,* somewhat inadequately translated as anxiety. It is not fear, for it does not have an object. It is, rather, the awareness of one's finitude, the awesomeness of one's lonely sojourn in the world. With the antiseptic precision of the surgeon's knife, it cuts the ties of security and at the same time opens up the vistas of human possibilities. *Angst* makes possible authentic existence, for it confronts us with our finitude and the possibility of a freedom which annihilates bondage to worldly care.

For the existentialist "being-in-the-world" is "being-with-others." Worldly care enslaves us because it transforms others into objects. In such a way our relationships with others become depersonalized and defined by manipulation and alienation. Depersonalization on a mass scale leads to the mass society in which freedom is jeopardized.

Our homelessness in the world is especially highlighted by chance, guilt, and death. In particular this is true of death. Death is the one inescapable reality of human existence. We cannot escape from it, although we can, of course, suppress our consciousness of it. But to live as though death is not a reality is to live inauthentically. Only when the reality of death is faced squarely can we be liberated from its crippling tyranny; only then is the present given to us to make our own. Once the present is our own, the texture of our lives is enriched, for it gains a true decisiveness and significance.

It is quite understandable, therefore, that Bultmann would see in existentialism the categories through which to present the Christian message. Bultmann accepted, for example, that the teaching of Jesus was indeed centred upon the Kingdom of God, as Weiss had insisted. But Bultmann reinterpreted thoroughgoing eschatology: "The Kingdom of God is a power *which, although it is entirely in the future, wholly determines the present.*"[87] Bultmann's explanation of what he meant by this is a very clear statement of his existential interpretation:

> It determines the present because it now compels man to decision; he is determined thereby either in this direction or in that, as chosen or as rejected, in his entire present existence. . . . The coming of the Kingdom of God is therefore not really an event in the course of time, which is due to occur sometime and toward which man can either take a definite attitude or hold himself neutral. Before he takes any attitude he is already constrained to make his choice, and therefore he must understand that just this necessity of decision constitutes the essential part of his human nature. . . . If men are standing in the crisis of decision, and if precisely this crisis is the essential characteristic of their humanity, then every hour is the last hour, and we can understand that for Jesus the whole contemporary mythology is pressed into the service of this conception of human existence. Thus he understood and proclaimed his hour as the last hour.[88]

For Bultmann, the action of God in the world must perforce be existential, for God is not perceived objectively, nor can God's action be verified empirically. The modern scientific view of the world, in fact, militates against the perception of God acting in the world through supernatural events.

> In mythological thinking the action of God, whether in nature, history, human fortune, or the inner life of the soul, is understood as an action that intervenes between the natural, or historical or psychological course of events; it breaks and links them at the same time. The divine causality is inserted as a link in the chain of events which follow one another according to the causal nexus. . . .[89]
>
> Modern science does not believe that the course of nature can be interrupted or, so to speak, perforated, by supernatural powers.[90]

Only the person of faith perceives God acting in the world; it is hidden from others. The person of faith sees God's action not as something which "happens between the worldly actions or events but as [something] happening within them."[91] When the believer sees God acting in an event, he or she cannot present the meaning of that event to another and expect the same kind of appropriation. God's action is hidden and paradoxical because the believer sees it where others would not. There is a striking example of such faith in the life of Henning von Tresckow, the man who put the bomb on Hitler's airplane. When the attempt on Hitler's life failed, Tresckow committed suicide. In spite of the apparent eclipse of the power of God in this dark moment, Tresckow made a moving affirmation of faith before he died.[92]

The supreme decisive act of God was in Christ. The person of faith is the one who responds to the proclamation (the kerygma) of the Christ event. God addresses himself to us in the kerygma. Bultmann is at one with Paul in seeing this central affirmation of the New Testament as a scandal and a stumbling block. Bultmann has no intention of removing this stumbling block, but he does wish to remove another: the mythological world-view of the New Testament.

The New Testament reflects a mythological view of the world foreign to our modern understanding of it. Jesus is presented as a divine, incarnated, pre-existent being who atoned for human sin with his own blood. He was resurrected from the dead, ascended into heaven, and will return again (in the imminent future) on the clouds and bring history to an end. As every schoolboy knows, says Bultmann, history did not come to an end.[93] Moreover, a "corpse cannot come back to life or rise from the grave."[94] Bultmann believes that he speaks for "modern man" when he says, "An historical fact which involves a resurrection from the dead is utterly inconceivable."[95] But it is not just the miracles of the New Testament which are problematic, but its whole cosmological framework:

> In the middle is the earth; above it is heaven, below it is the subterranean world. Heaven is the dwelling-place of God and of the Celestial beings, the angels; the lower world is hell, the place of torment. But the earth itself is not simply the scene of natural everyday events, of forethought and of labour, in which it is possible to reckon with a regular and unchanging order; this earth too is the scene of the action of supernatural forces, of God and of his angels. These supernatural forces intervene in natural events, in the thoughts, in the will, in the actions of men.[96]

To say that the world-view of the New Testament is mythological is not, however, to say that the New Testament has nothing to say to us today. Myth is a way of expressing deep truths about our existence.[97] The task of the New Testament exegete is to translate these myths into concepts meaningful to us today or, to use Bultmann's term, to demythologize. Actually, as has been pointed out many times before, *re*mythologize would have been a much better choice of word. For what Bultmann attempted to do was not to delete the mythology of the New Testament but to translate it into existential terms, terms which would be more meaningful for the modern era. The task is to *understand the meaning of existence* which the myths embody. This process of demythologizing actually began in the New Testament itself with the theology of Paul and John. The importance of Paul and John for Bultmann's thought can be easily seen in the way he arranges his *Theology of the New Testament.*[98] The book is divided into four parts: (1) Presuppositions and Motifs of New Testament Theology; (2) The Theology of Paul; (3) The Theology of the Gospel of John and the Johannine Epistles; (4) The Development toward the Ancient Church. A chiastic arrangement can be discerned: ABBA. A is primarily historical, while B is primarily theological and normative.[99] Paul and John are normative because they exemplify the demythologizing program which Bultmann was merely continuing.

> The decisive step was taken when Paul declared that the turning point from the old world to the new was not a matter of the future but did take place in the coming of Jesus Christ. . . . To be sure, Paul still expected the end of the world as a cosmic drama . . . but with the resurrection of Christ the decisive event had already happened . . . . After Paul, John de-mythologized the eschatology in a radical manner. For John the coming and departing of Jesus is the eschatological event; . . . the resurrection of Jesus, Pentecost and the *parousia* of Jesus are one and the same event, and those who believe already have eternal life.[100]

Bultmann believed, then, that the precedent for his demythologizing had been set by the New Testament itself. But how are we to go about this demythologizing? Here we light upon a fundamental principle of Bultmann's hermeneutics. Rejecting "the principle of the empty head"[101] (presuppositionless interpretation), Bultmann says that to interpret a text we must attempt to come to it not without presuppositions but rather with the right presuppositions. Our presuppositions determine the questions we ask. If we think the historian's task is to cull facts out of the past, then the questions we ask of history will be determined by that presupposition. We can come to the New Testament with all kinds of questions -- what kind of sociological hierarchy is implied by it, what kind of agriculture was practised at the time, etc. But only existential questions will get to the heart of the New Testament. The New Testament is about human existence.[102] By treating the New Testament existentially, such terms as sin, spirit, flesh, death, and freedom take on a new relevance applicable to the person of today. It should be stressed that although Bultmann is using existential categories to understand the New Testament, ultimately he differs from someone like Heidegger in his religious commitment. Existentialism is in and of itself only a way of analyzing the world; deliverance from its bondage comes through God's act.[103]

It is at this point we note the essentially religious stance of Bultmann. Many Christians have criticized him for being philosophical and not religious. Barth, for example, says that Bultmann errs in his use of existentialism to interpret the New Testament because the New Testament is not about human existence, but about the self-revelation of God.[104] Others have criticized Bultmann for being religious and not philosophical and for not carrying to its logical conclusion his demythologizing programme. He does not, for example, demythologize the action of God in Jesus Christ.[105] This latter criticism is the weightier one, for it does seem somewhat paradoxical that Bultmann has no qualms about demythologizing the cross and the resurrection and yet baulks at demythologizing the action of God in Christ. Bultmann was aware of the criticism and defended himself thus:

> There certainly are . . . those who regard all language about an act of God or of a decisive, eschatological event as mythological. But this is not mythology in the traditional sense, not the kind of mythology which has become antiquated with the decay of the mythical world-view. For the redemption of which we have spoken is not a miraculous supernatural event, but an historical event wrought out in time and space.[106]

For many, however, this comment of Bultmann's did not resolve what they saw as a fundamental dilemma in Bultmann's thought. He accepts the decisive action of God in Jesus Christ -- an historical event -- but maintains that Christian faith requires no more that the "that" of Jesus. This disquiet with Bultmann's position eventually led some of his own followers to take issue with him. In 1953, in a now famous address, Ernst Kaesemann called for a new quest of the historical Jesus.[107] He argued that it was important to ascertain whether the teaching of the early Church was in continuity with that of Jesus. The "old quest" erred in its assumption that we could write the "story" of Jesus. This cannot be done. But the New Testament scholar -- *qua* historian -- has to acknowledge that some pieces of Synoptic tradition are historical. The Gospels do not simply tell the story of Jesus, but neither are they just the story of the pre-existent and exalted Lord.

Bultmann replied to Kaesemann by insisting that there was continuity, but that the continuity was between the historical Jesus and the primitive Christian kerygma, not between the historical Jesus and the Christ.

> The Christ of the kerygma is not a historical figure which could enjoy continuity with the historical Jesus. The kerygma which proclaimed him is a historical phenomenon, however. Therefore it is only the continuity between the historical Jesus and the kerygma which is involved.[108]

Bultmann reiterated his position that the kerygma presupposes the historical Jesus but does not show an interest in the content and character of his teaching.

Despite Bultmann's reply, Kaesemann's address was the signal for many of Bultmann's followers to begin the "new quest of the historical Jesus."[109] This new quest does not seek the factual (*historisch*) Jesus but the interpreted historical (*geschichtlich*) Jesus of the kerygma. The best book to come from this school is Guenther Bornkamm's *Jesus of Nazareth.*[110]

In summary, we have seen that Bultmann was most concerned to preserve the integrity of faith. The scientific world-view of people today makes it impossible for them to accept the mythology of the New Testament. Bultmann therefore embarked upon a non-recognitive interpretation of the New Testament, using existential categories. For him, faith does not intend past particulars -- the "that" of Jesus is presupposed, but nothing beyond this. He therefore repudiated the quest of the historical Jesus. Over against Bultmann there are those who argue that it is important to capture the conscious intentions of the early Christians and of Jesus himself; otherwise, there is a lack of continuity in the Christian tradition, and unless caution is exercised, Christianity becomes a timeless truth or "symbolized principle" not anchored in historical revelation. The Christian message is not a summons to a speculative possibility of existence; it points to Christ, who existed in the actualities of history.

5. *The Mysterious Christ*

What has our survey of these four major figures in the history of New Testament interpretation revealed?

The significance of Reimarus's work lies in the fact that he challenged in a systematic way the idea that the Gospels give us an accurate account of the historical Jesus. His own attempt to give an account of the life of Jesus (Jesus was really a revolutionary who failed) has not been widely accepted. Nevertheless, the quest of the historical Jesus really began with Reimarus, for after him all scholars in the quest had to deal with the question of the historicity of the Gospels.

Reimarus did not challenge the view that the Gospels *intended* history. What he challenged was their veracity, not their historical intent. It was Strauss who challenged the idea that the Gospels intended history. Strauss classified the Gospels as mythical. Thus, Strauss shifted the ground of debate from the question of whether the history contained in the Gospels was accurate to the question of whether the Gospels belonged to the genre of history. The great merit of both Reimarus and Strauss was that they widened the horizon of New Testament scholarship by posing new questions in a thoroughly analytical way.

Schweitzer, building upon the work of Weiss, challenged previously held conceptions about the Kingdom of God. The Liberals were wrong in reducing the Kingdom of God to a spiritual and ethical concept.[111] The Kingdom of God is not "within you" (Lk 17:21), rather, it is supernatural and catastrophic and comes at the end of time by God's initiative alone. The world in which Jesus lived -- dominated as it was by eschatology -- is thus essentially foreign to us today. The question became: If the world of Jesus is so foreign to us, how can his message be meaningful for today? Bultmann tried to answer this by translating Jesus' message into existential terms. Jesus taught that, as the end of the world was near, Israel was summoned to urgent decision. So it is with each individual -- he or she faces death and is therefore summoned to urgent decision. We can languish in inauthentic existence,[112] or we can affirm life and live authentically.[113]

Reimarus and Schweitzer represent one approach to the history of Jesus, and Strauss and Bultmann another. There are those who follow Reimarus and Schweitzer and attempt an historical reconstruction of Jesus from the Gospels,[114] and there are those who follow

Strauss and Bultmann and see the historical Jesus as either inaccessible or unimportant.[115] Each of these approaches has elements of validity. It is important to attempt to ascertain what Jesus said and did, and we cannot totally disregard the historical Jesus. If, for example, Reimarus were right, and Jesus was a revolutionary who failed, it would surely damage the integrity of the Christian faith. In this sense Reimarus's reconstruction of the historical Jesus is potentially destructive of the faith and quite different from that of Schweitzer. Schweitzer's picture of Jesus is not that of a deluded fanatic,[116] but rather that of a figure whose message was formulated within a world-view which is foreign to us. The question Reimarus poses for us is whether Christianity is derived from a fraudulent presentation of the history of Jesus. The question Schweitzer poses for us is whether we can bridge the hermeneutical gulf between the beliefs and expectations of Jesus and his contemporaries and those of the modern world. Schweitzer's Jesus is no failed revolutionary, but a sublime and heroic figure whose will and authority traverse the ages.

Christians, then, should not disregard the historical approach to the life of Jesus. Strauss and Bultmann have done a great service, however, in questioning the appropriateness of the purely historical approach. To turn Christianity into a kind of Jesusology which is dependent upon the whims and uncertainties of historical reconstruction for its verisimilitude is to misconstrue its nature. In each age new questions are posed and different scholars bring differing presuppositions to the historical task. There are thus bound to be differing interpretations of Jesus. The fact that there are differing interpretations should not, however, be disconcerting to the Christian. Even if we could recount definitively and accurately the life of Jesus on earth, such an account would not exhaust his meaning,[117] for Jesus as the risen Lord is irreducible to merely human descriptions of him. On the one hand Jesus is the historical figure who lived two thousand years ago in Palestine. On the other hand he is the risen Lord who is present in his Church today and who works through it, ever sustaining and renewing it. He is both an historical figure and a transhistorical saviour, and this paradox lies at the very heart of Christianity. Throughout the history of the Church there have always been oscillating tendencies to stress, on the one hand, the humanity and historicity of Jesus and, on the other hand, his divinity and transhistorical dimension.[118] But in the Church at large the temptation to deny one in favour of the other has always been resisted. Grappling with the paradox of the Divine Man gives Christianity a creativity which makes it a vital and dynamic faith. The attempt to plumb the depths of the person of Christ is a process which constantly confronts the believer with new questions and new perspectives. In one age

Jesus may be comprehended best in the mode of apocalyptic, in another in the mode of existentialism. But he is never fully grasped -- much less exhausted -- by any mode. He is essentially an elusive and mysterious figure. There is no finer description of this mysterious Christ than that which concludes Schweitzer's *Quest*:

> He comes to us as One unknown, without a name, as of old, by the lakeside, He came to those men who knew Him not. He speaks to us the same word: "Follow thou me!" and sets us to the tasks which he has to fulfil for our time. He commands. And to those who obey Him, whether they be wise or simple, He will reveal Himself in the toils, the conflicts, the sufferings which they shall pass through in His fellowship, and, as an ineffable mystery, they shall learn in their own experience Who He is.[119]

Chapter Two

The Development:

Belief and Practice in the Early Church

We concluded our last chapter by suggesting that the Lord Jesus is essentially a mysterious figure who is never fully grasped by human descriptions of him.[1] A purely historical interpretation of him is, we maintained, essentially inadequate. In early Christianity, to use Bultmann's well-known aphorism, "the proclaimer became the proclaimed."[2] "If," says Paul, "you confess with your lips that Jesus is Lord, and believe in your heart that God raised him from the dead, you will be saved" (Rom 10:9). Jesus as risen Lord is the focus of Christianity.

If, then, a purely historical investigation into the story of Jesus does not enable us to grasp fully the essence of Christianity, what other fruitful avenue of approach is there? The historical approach must be complemented by an approach which will yield a better understanding of what it means to confess Jesus as Lord. It is at this point that we come upon the phenomenon of the early Church, for it is within it that we find the exploration of the significance of the person and saving activity of Jesus.[3] These explorations were given concrete formulation at the Great Councils of the third and fourth centuries and became normative articles of faith. It is seductively tempting to take these articles of faith and seize upon them as expressions of the essence of Christianity. Such an approach has many drawbacks, not the least of which is that it ignores an important debate about the nature of early Christianity. There are those who argue that the normative or orthodox articles of faith promulgated at the Great Councils are not, in fact, good indicators of belief and practice in the early Church.

One cannot have heretical belief without a norm against which it is judged to be false or defective; that is, orthodoxy will only reach formal definition with the appearance of heresy. This observation throws light upon the so-called Eusebian view of history, which conceived of orthodoxy as *historically* prior to heresy. This classical pattern of the development of early Christianity runs unbelief, right belief, deviations into wrong belief. First, unbelievers were converted into Christian believers, and only later were the apostles commissioned to take this unadulterated gospel to the portions of the world allotted to them. It was only after the death of the apostles that heresy crept into the Church.

It was exactly this schematization of the development of early Christianity which Walter Bauer criticized in his important book *Orthodoxy and Heresy in Earliest Christianity.*[4] The book occasioned a continuing debate[5] which directly touches upon our inquiry. Bauer's view of early Christianity was that it was marked by fluidity and confusion. What was later labelled "orthodox" came *after* heresy in many areas; the victory of what we now call orthodox was due almost entirely to the influence of the Roman Church. If Bauer's view of early Christianity is essentially correct, it would have considerable implications for our inquiry.[6] Accordingly, it is necessary to discuss in some detail belief and practice in early Christianity.

The most important work on this question is H. E. W. Turner's *The Pattern of Christian Truth: A Study in the Relations Between Orthodoxy and Heresy in the Early Church.*[7] He begins by describing the classical theory of orthodoxy and heresy. Heresy is not present before true doctrine. But the heretics themselves challenged such an orthodox caricature. Marcion himself claimed to be a conservative and certainly not an innovator. The Gnostics claimed to represent a secret tradition no less authentic than that of the Church. And heretics no less than orthodox argued from scripture and laid claim to the title "Christian."[8]

Turner maintains that the classical view will not stand -- not least because it supposes a static conception of orthodoxy. Heresy is certainly not a deviation from a fixed and static norm. The Apostolic Fathers cannot be reduced to a single doctrinal common denominator; nor can the New Testament be embraced under a single theological rubric.[9]

Turner briefly reviews the three modern alternatives to the classical view of orthodoxy offered by A. Harnack, M. Werner, and R. Bultmann. Each of these expositors stresses the diversity and fluidity of early Christian thought, in opposition to the notion of fixed and stable norms. Each seems to suggest that the "orthodoxy" which was eventually victorious was discontinuous from the original Christian faith.

Turner finds himself out of sympathy with extreme views on the orthodoxy/heresy issue. The classical notion of a fixed and static doctrinal norm is too simple. On the other hand, the view that sees the resultant victorious faith as a travesty of its former self is too severe. These alternative modern views imply too high a degree of openness or flexibility. Accordingly, Turner sets himself the task of bridging the gap between the two views:

> The development of Christian theology as a whole (and not merely in the Patristic period) may be perhaps better interpreted as the interaction of fixed and flexible elements, both of which are equally necessary for the determination of Christian truth in the setting of a particular age.[10]

What are the "fixed elements" in the Christian tradition? First, says Turner, there are the "religious facts themselves, without which there would be no grounds for its existence."[11] This is a fundamental point for Turner: "The Church's grasp on the religious facts was prior to any attempt to work them into a coherent whole."[12] Turner gives the name *lex orandi* to the notion of the "the relatively full and fixed experimental grasp of what was involved in being a Christian."[13] Thus, for instance, Turner maintains that Christians lived trinitarily long before the evolution of Nicene orthodoxy.

Further elements of fixity lay in the biblical revelation, the creed, and the rule of faith. There is a direct sequence from the New Testament kerygma through the stylized summaries of *credenda* to the earliest credal forms themselves.[14] The creeds may well mark the beginning of a new stage. Yet this stage itself is a continuation "of a process which takes its origins from the formalized oral tradition of the Apostolic Church itself."[15]

Among the flexible elements within Christian thought, Turner finds one of the most important to be "differences in Christian idiom." Many see a radical difference between an eschatological and metaphysical interpretation of Christianity. Turner, however, maintains that the "Christian deposit of faith is not wedded irrevocably to either idiom but is capable of expression both ontologically and eschatologically."[16]

> The selection of a distinctive theological idiom, whether it be eschatology, ontology, or even in more recent times existentialism, illustrates one possible element of flexibility in Christian thinking.[17]

Flexible elements also lie in the individual personalities of the theologians themselves.

Turner's main contention is that the situation described by Bauer is more adequately explained by the existence of a "penumbra" or fringe between orthodoxy and heresy; the line of division between the two was not nearly as sharp as Bauer avers. Bauer's treatment is vitiated by his failure to attain an adequate view of orthodoxy: he does not allow for its richness and variety. In short, "orthodoxy resembles not so much a stream as a sea, not a single melodic theme but a rich and varied harmony, not a single closed system but a rich manifold of thought and life."[18]

Turner's critique of Bauer is captured in the following excerpt:

> His fatal weakness appears to be a persistent tendency to over-simplify problems, combined with the ruthless treatment of such evidence as fails to support his case. It is very doubtful whether all sources of trouble in the early Church can be reduced to a set of variations on a single theme. Nor is it likely that orthodoxy itself evolved in a uniform pattern, though at different speeds in the main centres of the universal Church. The formula "splinter movement, external inspiration or assistance, domination of the whole Church by its orthodox elements, tributes of gratitude to those who assisted in its development" represents too neat a generalization to fit the facts. History seldom unfolds itself in so orderly a fashion.[19]

In his theological analysis of heresy, Turner tests the claim of various heresies to be Christian. If we put the question thus -- how do the various heresies deviate from the norm? -- Turner answers that Gnosticism is a dilution of Christianity by "alien elements"; Marcionism is a "truncation"; Montanism is a "distortion"; Arianism is an "evacuation." Those heresies which preserve the past without reference to the present Turner describes as "archaisms."[20]

Turner's effort is to show how heresy is not so much a questioning of the tradition[21] as a whole, but rather a failure to grasp the right relation of elements within the tradition. There is an intuitive rejection of heresy through Christian common sense. The results of such development must be tested not only by the principle of coherence, "the logical articulation of the Christian faith into a systematic whole, but also by the further principle of correspondence with the biblical facts themselves."[22]

Bauer and Turner are the two initiators of the discussion on orthodoxy and heresy in early Christianity, and their works have silhouetted quite sharply the issues at stake. The different conclusions of Bauer's and Turner's works arise from the differing perspectives and presuppositions they bring to their inquiry. It is to a careful examination of these presuppositions that we now turn, for upon examination these differing presuppositions are seen to be grounded in quite different approaches to the history of early Christianity.

1. Orthodoxy, Heresy, and Development: The Semantic Problem

One of the difficulties of dealing with Bauer's work is semantic. How should orthodoxy and heresy be defined? Bauer says that in using these terms he is referring to what one "customarily and usually understands them to mean."[23] Presumably he hoped thus to avoid confusion. In fact, his lack of a precise definition has created confusion.[24] The author of the second appendix to his book, pointing to the varying definitions of orthodoxy given by Moffatt, Ehrhardt, Turner, and Koester, asks: "Is there today *any* commonly accepted meaning of 'orthodoxy' such as Bauer wished to presuppose?"[25]

The matter of definition is indeed a confusing one. Georg Strecker, for example, argues against "dogmatically conditioned" definitions of orthodoxy and heresy on two grounds. First, such definitions are *a priori*; they do not arise "phenomenologically on the basis of statements by New Testament writers."[26] Secondly, such definitions are by nature incompatible with history, for they enshrine value judgments, and "value judgments," as Bauer himself intimates, "are not the business of the historian."[27] On the other hand, Strecker urges the necessity of entering into "the period and thought world" of the writers to be examined.[28] Yet he overlooks the fact that historians themselves (and not their "sources") pose the historical question. Historians may ask any question they want. Moreover, Strecker seems to presuppose that the question of whether there is a normative or distinctively self-conscious conception of Christian belief and practice in early Christianity has already been settled in the negative. Only detailed historical investigation would, in fact, prove such a case.

Not only is Strecker unpersuasive in his view that the historical question of a normative early Christian belief and practice has been settled, but he is also open to severe criticism concerning his second view, that having to do with value judgments and value-free judgments. We understand history to aim at settling matters of fact. But this cannot possibly exclude value judgments from the matters of fact to be settled. History does not say who (if anyone) was orthodox, who (if anyone) was heretical. It does, however, have something to say about who *claimed* to be orthodox and who charged whom with heresy. Again, it is a positivistic illusion to say that value judgments have no role in the work of the historian. They should not, indeed, serve as a substitute for evidence. They guide the choice of historical questions without presuming to answer them.[29]

In the light of these preliminary observations, what kind of definition of orthodoxy and heresy would be relevant to the thinking of the early Christians themselves? The early Christians' world of meaning belongs indispensably to the data of which the historian must take account. The definition ought, ideally, to be at once specific enough to meet real issues and flexible enough to apply to different times and places. It should serve us well in recovering horizons often less differentiated[30] than our own.

The call for this kind of definition is as old as Socrates. When Socrates asked "What is justice?" he was looking not for a list of the acts of a just man but for a proposition capturing the nature of justice, one applicable to every instance of justice and to nothing other than justice. Socrates was in quest of an insight not into the usages of words, but into the essences of things.

In the light of this reflection on definition, let us consider Bauer's critique. Bauer's starting point was the doctrinal commitments that emerged dominant in Christianity by the end of the third century. The commitments were specific articles of faith. The claim was that these articles had defined Christian faith from the beginning. Taking "orthodoxy" to signify the commitments, Bauer set out to put the claim to the test of history -- and found it wanting.

Now this project, straightforward as it might appear, has two related drawbacks. First, it skipped too quickly over the thought of the participants in the ancient history of orthodoxy and heresy. Secondly, it did so because in his definition of orthodoxy Bauer took precisely the tack condemned by Socrates as inadequate. Like the inept Athenian who defined not the nature of justice but the acts of a just man, Bauer settled on the material components of orthodoxy in the third and fourth centuries. He saw that the claim that these material components of orthodoxy were present in first-century Christianity was wrong. But what he himself failed to do was to settle on a heuristic[31] definition of orthodoxy; that is, he did not offer a formal definition which, as an invariant structure, could take account of *development*. He therefore lacked the conceptual tools to deal with orthodoxy as a development incorporating the past, accommodating the present, and anticipating the future. Indeed, throughout the book we have the impression of entrenched rival factions waging war for souls from clearly defined positions. Hence, it is with some apparent surprise that Bauer notes:

> The religious discussion which brought about the split in Rome between Marcion and orthodoxy was of a special sort. At least at the outset, it was not thought of as a struggle for the souls of Roman Christians fought from already

established positions, but as an effort to ascertain what the true meaning and content of the Christian religion really is, and to that extent it was somewhat comparable to the apostolic council (Acts 15).[32]

As will be seen shortly, for the early Christians, revelation was the major presupposition of orthodoxy, which was conceived of as that which responds to and is grounded in divine revelation. As such, it is a formal definition, the material components of which are provided by the relevant data of history. It is heuristic in that it enables us to label the intended unknown and to set down all that can be said about it in a particular historical age under investigation.

2. A Point of Departure: The Lex Orandi

Turner's attempt to reconstruct the history of early Christian thought is undoubtedly more accurate than that of Bauer. Bauer, because he conducted his inquiry without reference to any formal principle of orthodoxy, offered a view of early Christianity which suggested a more differentiated mentality than we can reasonably suppose was there.

Like Bauer, Turner concerned himself with the doctrinal content of orthodoxy and heresy. But he did so in a somewhat different way. As we have seen, he argues that an examination of the self-understanding of post-apostolic Christianity revealed a healthy interaction between fixed and flexible elements. Only by grasping this point does the unity and diversity of early Christianity become intelligible. Moreover, if we take account of the mentality of the earliest Church, we cannot but come upon the phenomenon Turner calls the *lex orandi*; that is, the consciousness of standing in a faith relationship, a response to divine revelation essentially consisting in thanksgiving for the boon of salvation. We infer from the performance (*Vollzug*) of Christian faith that it is at once an exigence for and a resource of theology.

The more recent debate reflects a marked lack of sympathy with Turner. Koester complains that Turner's notion of *lex orandi* is "theologically mute."[33] Ehrhardt thinks that Turner retreats into dogmatic categories and implies that he has missed the point.[34] To facilitate further discussion, it is thus necessary at this point to endeavour to uncover some of the root presuppositions which separate Turner from many of the other participants in the debate.

There is an obvious difference between Turner and Koester over the concept of *lex orandi*. It is not entirely clear in what sense Koester thinks that Turner's *lex orandi* is "theologically mute."[35] If he means that it is theologically unexplored or not fully formulated, then Turner would doubtless agree, for doctrinal explorations and formulations are the province of the *lex credendi*. To grasp this point one must remember that *our* horizons are genetically[36] different from those undifferentiated horizons of early Christianity.

Let us pursue this tack a little further by examining early Christian thought in the light of the *lex orandi*. The formulation of doctrine was not the primary purpose of the early Church; rather, it saw itself as charged with the mandate of preaching the gospel and concomitant teaching of the Christian life. It was in the course of this mission that "intellectual scaffolding" was erected around the living realities grounded in Christian experience mediated through the Church. The communication of these realities self-evidently implied conceptualization. But the need to present Christianity in an acceptable form -- implying, as it does, the use of a conceptual framework and linguistic idiom -- arose primarily in the contexts of propaganda and apologetics. The Church needed conceptual tools in order to be missionary, and sheer survival dictated their use in the rebuttal of pagan criticism. The first attempts at the articulation of the Christian faith were not undertaken for their own sake; Origen's *First Principles* is the first treatise on theology conceived out of an exclusive passion for theologizing.

> Thus the Biblical data are mediated through the medium of the *lex orandi* of the Church. All the major doctrines of orthodoxy were lived devotionally as part of the corporate experience of the Church before their theological developments became a matter of urgent necessity.[37]

The *lex orandi* preceded the *lex credendi*. Thus, as Turner and others have noted, trinitarian religion preceded and necessitated trinitarian theology. The triune baptismal formula of Mt 28:19 is universally cited in the practice of the early Church. Even the Arians accepted the text. The doctrine of the trinity was primarily an extension of and exploration into the baptismal formula. It was this *lex orandi* basis which led the early Church to expand instinctively the Old Testament dictum "As the Lord liveth" into "As the Father liveth, as the Son liveth, as the Holy Spirit liveth."

Lex orandi, then, covers the instinctively taken up devotional and liturgical attitude of the early Church. It harmonizes with the close connection between spirituality and theology and throws light on otherwise seemingly intractable problems. In this sense it is akin to the

Russian orthodox term *sobornost* -- "togetherness," the devotion of the faithful as a springboard and control of theological speculation.[38]

As we have said, missionary and apologetic motives dictated the articulation of Christian faith. The degree and precision of such articulation was dictated by historical circumstance. The Church was subject not only to attack from without but also to disruption by differences from within,[39] as certain tendencies were seen as leading to positions which were irreconcilable with the bases of the faith. Historically, it is difficult to see how it could have happened otherwise. It is wrong to think that the Church had a blueprint from the beginning which was the touchstone of correct Christian belief, for the early Christian consciousness was relatively undifferentiated. Heresy was not a known in the sense of the transgression of a fixed theological law.

In the beginning the Church was a collection of people who were bound together by the common belief that Jesus was the Christ, the bringer of salvation. The exact implications of such belief had to be worked out in the course of time as the need arose. The history of the early Church is thus a history of doctrinal explorations; it was not immediately obvious that certain avenues were culs-de-sac. Yet, although the journey down a certain false avenue may have begun, once the life and reflection of the Church had revealed that route to be a dead end, it was abandoned and others were explored. Those who continued along that road eventually reached the point of no return, and such the Church had to disown as it carried on its search for truth down other avenues.

The first major problem the Church faced in erecting its intellectual scaffolding around the faith was the reconciliation of its trinitarian theology (given in the *lex orandi*) with the monotheism of the faith from which it sprang. This occupied the mind of the Church until Nicaea. The relation of the Father to the Son was the first item of reflection, as Christ was so central in preaching and teaching. Once this was settled, the precise nature and position of the Holy Spirit within the Godhead was formulated -- as is evidenced by the quite rapid development of doctrine of the Holy Spirit in the period from A.D. 327-381.

The formulation of Christian doctrine was the result of the interplay of various ways of thinking within the Church. The characteristics of the whole Church militated against extremism. Even the great individual traditions of the Church tended mutually to correct each other. The West, with its love of the concrete and the balanced, was a good foil for the East, with its love of the mystical and speculative.[40] It is quite valid to see the development of early Christian doctrine as an interaction between fixed and flexible elements. It is, of course, wrong to envisage this development as having fixed and narrow limits:

> The customary limitations imposed by human sin, human error, and human blindness can be observed even here. Christian theology is not exempted from the law of oscillation which applies to all branches of human thought. Premature syntheses required subsequent modification and the dangers of distortion and accretion were not slow in making their presence felt.[41]

If the progression of orthodoxy did not proceed along the "straight and narrow" in quite the way envisaged by the classical approach, neither is it true to say that it often wandered off the road completely. To be sure, the development of early Christian doctrine is characterized by oscillations, but the quest for balance was always there, with the givenness of the New Testament data as the fulcrum.[42]

The development of Christian thought begins with faith data; from this givenness certain inferences are drawn as historical circumstances dictate. These inferences at a later stage -- after they have become explored and tested -- become presuppositions. Thus, as Dodd points out,[43] the content of the kerygma entered the rule of faith which second- and third-century Christian theologizing recognized as the presuppositions of faith. Out of the rule of faith the creeds eventually emerged. The doctrinal formulations of the creeds were tested by their coherence[44] and correspondence to the early Christian faith data.

3. "Development" Revisited

Turner is surely right in seeing the *lex orandi* as crucial to the understanding of early Christian thought. But the differences between Turner and others are not confined only to the *lex orandi*; upon close scrutiny another vital difference emerges. While such as Koester stress the syncretistic nature of early Christianity,[45] Turner stresses the identity of subject matter and congruity of content but sees a difference of style, idiom, and historical context. That is, Turner argues for a *dynamic*[46] unity of Christian development.

Turner believes that the Church's rejection of Gnosticism indicates that Christianity is not intrinsically a syncretism.[47] He admits that the Church used what it thought to be the best available categories, usually drawn from Greek philosophy, which were not designed with the Christian realities in view. But Turner considers the real question to have been whether these realities would be "evacuated," "diluted," or "truncated," or whether the thought-forms could be successfully adapted to their new purposes.

Yet, although Turner sees the development of early Christian doctrine as an interaction between "fixed and flexible" elements, as we have seen, he does stress that it is wrong to envisage this development as having fixed and narrow limits. In this connection, it must be acknowledged that development is mysterious. As B. F. Meyer says:

> Development, unlike organic growth, unlike logical deduction, takes place in the sphere of spirit, subjectivity, freedom, meaning, and history. It is unpredictable. Its authenticity is not discerned equally by all, nor all at once. It is taken in piecemeal, by a learning process, and is satisfactorily grasped after the fact.[48]

If we think of orthodoxy as a moving norm, then as normative, past orthodoxy is a discernible and powerful ocean current; as moving, present orthodoxy is mysterious and a challenge to discernment.

So, then, Turner supposes that Christianity has a distinctive identity, that *das Wesen des Christentums* is a meaningful phrase, that there is such a thing as the "Christian deposit of faith." Koester and others seem to make no such supposition. On the contrary, they seem to suppose the opposite: that Christianity has no intrinsic substantial identity, that it is a syncretism or, more exactly, an ongoing multiplicity of interpretations with family resemblances.[49]

Turner's starting point is the givenness of God.[50] This is perhaps a point worth dwelling on. Christianity saw itself as being born not of its own act but of God's. Despite the diverse perspectives of New Testament writers, each has the conviction that the selfhood of Christianity was born "from above" and guided by God Himself.[51] Christians never thought of their existence as auto-justified; rather, they held their existence in the consciousness of a divine mandate.

Paul certainly conceived of the gospel as having a distinct identity. Right belief was that which mediated salvation. But salvation is God's gift (I Thess 2:13; Rom 1:16); salvation is contingent upon the truly revealed. It is this lived consciousness of the interdependence of salvation and the truly revealed that is central to our discussion.

When Turner speaks of the *lex orandi* which is to be conceived of as *pre*theological,[52] he is speaking of the heuristic effort to grasp the givenness of God -- the "religious facts." There is the "that which" which is experimentally grasped, the experimental grasp, and the reflective thematization of the experimentally grasped. No concrete definition of what it is to be a Christian can be comprehensive. In this sense the "that which" which is grasped is ineffable. Obviously, there is a differentiation of consciousness between a Christian of the first century and one at Chalcedon. But these differentiations must be viewed in relation to what it is that is being expressed.[53]

Turner's "religious facts" are irreducible to merely human conceptions and descriptions and so may appear now in the mode of apocalyptic, now in the mode of metaphysics, now in the mode of existentialism, never fully grasped -- much less exhausted -- by any mode.[54] It seems that it is perhaps here that the widest gulf between the universes of discourse of Turner and others is located. Contrast this style of thought, for instance, with the notion that messiahship refers not to a role willed by God as a feature of his salvific plan but merely to our idea of such a role. It is true that one cannot say that Jesus is messiah on historical grounds, whereas one can (perhaps) investigate Jesus' messianic consciousness. But this latter procedure leaves open the decisive question of the status in reality of "the thing" (*die Sache*) referred to. As messiahship may thus be reduced to a mere idea having no reality *extra nos*, so "orthodoxy" may be reduced to an essentially empty claim. In both instances the reduction may take place without being argued for and so may function as an unexamined presupposition.

The Christian revelation was eschatological and definitive. Christians can never free themselves from this historical concrete revelation of God through Jesus Christ:

> Revelation is not . . . a set of principles and development a set of inferences, nor is revelation a "seed" and development its organic growth. Revelation is not a Christology, it is Christ. Growth, as understood in Ephesians, is the deepening, not of the knowledge of theology, but of "the knowledge of the Son of God"(4:13); the "inexhaustible riches" (3:8) are not riches of doctrine, but riches of Christ. Likewise, "the divine mystery in which all treasures of wisdom and knowledge are to be found" (Col 2:3) is precisely Christ himself.[55]

Thus, the hermeneutical questions which are so central to the discussion of the orthodoxy/heresy debate might well begin with the question of Christology and of Christ. Is Christology the derivative of "Jesus-in-his-Christological-integrity?" Or is Christology an imposition on "Jesus-as-man-like-any-other?" The first option affirms a *Sache* in which Christological *Sprache* is grounded: Jesus has the *extra nos* reality of "being the Christ." This is precisely the option supposed by early Christianity in accord with what Turner calls the *lex orandi.* Without it, the Christological struggles of the ancient Church from the Gnostic controversy to Nicaea hardly make historical sense.

4. The Distinctively Christian in the New Testament

Thus far we have indicated that we consider Turner's notion of *lex orandi* to be an essential cornerstone in any reconstruction of early Christian thought. We have also attempted to explicate what we consider a root presupposition separating Turner from such as Koester: the former argues for an organic unity of early Christian development and the latter for an artificial unity. We shall now attempt to vindicate Turner's position on this latter point by reference to the New Testament itself.

In Paul's words, to be a Christian is to put on Christ (cf. I Cor 3:10f.). The experience of Christ is the guiding principle of the Christian's life.

> There is a unity in all these early Christian books which is powerful enough to absorb and subdue their differences, and that unity is to be found in a common religious relation to Christ, a common debt to him, a common sense that everything in the relations of God and man must be and is determined by him.[56]

This experience is accessible through its manifestation in the life of the community. The cumulative self-understanding of the early Church community served as a regulative and stabilizing expression of this internalized "norm." As the Christian self-understanding was essentially ecclesial from the start, it was protected against arbitrary discontinuities and involved in the social process as a moving consensus. This notion of orthodoxy as a moving, organic[57] consensus evolving from implicit beliefs grounded in the experience of Christ seems to be in essence the same as that of Turner.

This assertion that the Christian community, the Church, of the first century was subject to self-regulation and correction from within is subject to the following objection: Was there a Church self-understanding in the first century? Or were there just autonomous, spontaneously founded churches? The answer is that the individual Christian communities understood themselves as particular realizations of the Church as one transcendental reality.[58] A community of Churches was a first-century fact, as the career of Paul illustrates.[59]

Moreover, the Church's self-understanding was controlled by *two* factors: not only the experience of Christ but also the scriptures. This latter point needs emphasis. The faith formulas of the early Church reflect a consciousness of continuity. The historic relationship with Israel was reflected in the covenant. The definitive character of the new covenant secured the present and the future; the past was secured by attestation of Jesus as *fulfilment* -- fulfilment of law, promise, prophesy, and type.[60] Paul, even if only in a limited sense, was conscious of this continuity.[61] His experience did not lead him to repudiate his heritage; rather, he saw it in an entirely new light.

We now turn to our fundamental and perhaps most contentious point that the Church did have an acute sense of the confessional principle.

> Had the Christians of the apostolic age not conceived of themselves as possessing a body of distinctive, consciously held beliefs, they would scarcely have separated themselves from Judaism and undertaken an immense programme of missionary expansion. . . . It was their faith in this gospel which called them into being, and which they felt obliged to communicate to newcomers. It would have been surprising if they had not given expression to it in their teaching as well as in their corporate life and organization. Like other religious groups with a saving message, they must have been driven by an inward impulse to embody it in their liturgy, their institutions, and their propaganda, and to seize every opportunity of harping on it.[62]

It is by specifically focusing on the *credal elements* in the New Testament that we hope to support our case. It is true that the New Testament will not support evidence for sacrosanct and stereotyped creeds. But evidence can be adduced for early faith formulas.

The New Testament abounds with evidence of an emphasis on the transmission of authoritative doctrine; for example, Jude vss. 3, 20; II Tim 1:13; 4:3; Tit 1:9; 1:13; I Tim 6:20; 1:19; 4:6; Hebs 3:1; 4:14; 10:23; 6:2. Moreover, this is hardly a process confined to the end of the first century, for Paul witnesses to a much earlier stage in its evolution; for example, Rom 6:17; 10:8ff.; I Cor 11:23-25; 12:3; 15:3-5.

The chief constituents in the preaching of this faith have been isolated in the classic study of C. H. Dodd, *The Apostolic Preaching and its Developments*.[63] However, this pioneering study does not give us the full picture by any means. For the emphasis in preaching was necessarily Christological; we must be aware of the basic suppositions that Christianity took over from Judaism -- for example, one God the

Father, maker of heaven and earth. We must also take into account the liturgical and catechetical life of the early Church. In baptism, for example, there was usually some sort of avowal of beliefs before the initiate entered fully into the Christian community. It is evident that such confessions would crystallize as a matter of course.

J. N. D. Kelly has collected the most impressive list of one-clause Christologies, bipartite- and tripartite-structured confessions, and concludes that it is manifest that there is a common doctrine shared by the Church from earliest times.[64] We will not repeat his arguments here. Rather, we will focus attention on just one faith formula, I Cor 15:3-5, to see if this sheds any light on our particular inquiry.

I Cor 15:3-5 is a confessional formula which Paul declares he had received and which he had handed on to the Corinthians:[65]

> That Christ died for our sins in accordance with
> the Scriptures
> And that he was buried
> And that he was raised on the third day in accordance with
> the Scriptures
> And that he appeared to Cephas, and then to the Twelve.

It is most probable that this kerygmatic formula was first composed in Aramaic.[66] As Paul says, he "received" the formula; it must therefore date from at least the forties of the first century. Moreover, it doubtless served catechetical purposes (the lapidary form of the formula shows it to be an entity in itself and not merely a list of topics).

What is the centre of gravity of this formula? The central affirmation is made in lines one and three; line two functions as a warrant for line one and similarly line four for line three. That is, the formula affirms that Christ died for our sins and was raised again on the third day. The emphasis falls on the motif of expiatory death and the resurrection; these motifs are explained as being "according to the scriptures." There was no need to emphasize that Jesus had died, but rather that he had died for *our* sins (not his own), and that moreover he had been raised from the dead. (These motifs are found in faith formulas elsewhere in Paul; for example, Rom 4:24f. and 8:34.) The last limb is important, for it locates the testimony which specifies faith. This last limb is not a "proof." What it shows is that early Christianity was committed to the apostolic tradition as the basis and rule of faith.[67]

The confessional formula in I Cor 15:3-5 and the ones similar to it illustrate convincingly the point we made at the beginning of this chapter: the proclamation of Jesus as risen Lord lies at the very heart of early Christianity. It is true that there is much dispute about whether the resurrection should be thought of in physical terms. Bultmann, it will be recalled, finds such an idea inconceivable. A discussion of this question is beyond the scope of this book. But leaving aside whether there was actually a physical resurrection or not, what is indisputable is that the early Christians perceived Jesus as a present experience. He was their Lord and Saviour, present in the worshipping community, sustaining and renewing the faith.

5. *The* Lex Orandi *and the Early Christian Style of Life*

What we have endeavoured to show is that in the New Testament we find faith data (in the kerygma and in the faith formulas) which demand a faith response. Given a faith response, the faith data (Turner's religious facts) are then proclamations within faith. That is, they function only within a faith perspective. What was it that animated, sustained, and nourished the early Church? From whence did it understand itself to derive its animating power? The answer to these questions is surely the key to understanding the life and development of Christianity, and it is in addressing precisely this question that Turner formulated his notion of *lex orandi.* The faith response is the response to the givenness of God. It is only within this context that one can begin to make sense of Christian development. The faith data are, of course, expressed in religious speech, and any use of language does involve the problem of the mediation of meaning. But religious speech is different from theological speech. It functions differently. Religious speech is the expression of worship; it is charged with thanks for the boon of salvation. Theology uses language to reflect often on the religious life, but more often on the character of the God who is worshipped. Perhaps a concrete example will illustrate the point. The theologian may say, "The Catholic faith is this: that we worship one God in Trinity, and Trinity in Unity; neither confounding the Persons, nor dividing the Substance." The initial response of the religiously committed person is more likely to be in the form of an utterance such as, "O Love that wilt not let me go."

It is because our technological society is so possessed with abstract and formal thinking that this kind of approach seems so foreign. Moreover, within the Christian tradition the stress on preaching, hearing, and understanding, so characteristic of Protestantism, has perhaps placed undue stress on what goes on in the mind. It has often been remarked, for instance, that the Old Testament contains very little actual "doctrine." As W. Eichrodt observes:

> Nowhere are formal "instructions" about the Being of God or his attributes delivered to the Israelite. His knowledge of God comes to him from the realities of his own life. He learns about the nature of God by reasoning *a posteriori* from the standards and usages of Law and Cult, which rule his personal life with divine authority, from the events of history and their interpretation by his spiritual leaders, in short, from his daily experience of the rule of God. By this means he comprehends the divine essence much more accurately than he would from any number of abstract concepts. The result is that the formation of such concepts in the OT lags far behind, while the same spiritual values which they are normally the means of conveying to us are yet uncompromisingly real and effective.[68]

As in the Old, so in the New. While the New Testament emphasis on "what is heard" (Rom 10:17) cannot, of course, be discounted, neither should theology be seen as an abstract intellectual "game" -- it arises out of the need to articulate prior religious commitment. Prayer in response to the boon of salvation is what Macquarrie calls "thankful thinking." Theology arises out of the whole life of faith; it cannot be isolated from it. "Theology itself, as the intellectual clarification and interpretation of faith, cannot be isolated from the whole life of faith. Theology makes sense only in the context of worship and action."[69]

As Macquarrie further points out, the knowledge of God cannot be described as either objective or subjective, because our knowledge of God is not like knowledge of things, nor is it like what we know of ourselves:

> We know God only because he lets himself be known, and therefore our knowledge of him is not the mastering, objectifying knowledge that is characteristic of natural sciences, but is a knowledge suffused with reverence and gratitude. The knowledge of God is inseparable from the adoration of God.[70]

A genuine theology, then, is shaped by a knowledge of God in prayer and worship.[71]

We have been specifically concerned in this chapter with Christian belief and practice in the early or classical period. We have made no attempt to trace Christian development through to modern times (such an enterprise would require several books). Many differing branches of Christianity have, of course, sprung up throughout its history, with quite significant differences between them. But all the different branches of Christianity, with varying degrees of enthusiasm, trace their origin back to the classical period and see it as normative for belief. Therefore, all attempts to describe or grasp the phenomenon of Christianity must focus on this period. We must now draw together the various strands of what we have said to form a conclusion about the phenomenon of Christianity.

Conclusion to Part I

Understanding the Phenomenon of Christianity

We began our discussion of the phenomenon of Christianity by saying that its unique focus was Jesus. In chapter one after our examination of four interpretations of Jesus we concluded that he was essentially a mysterious figure. We suggested in chapter two that in early Christianity belief was not deduced from abstract truths but emerged concretely out of the tension between intellectual reflection and the community's experience of Jesus as Lord. This, we argued, was a most fruitful avenue of approach in attempting to understand the phenomenon of Christianity. As was further pointed out in chapter two, we believe that there is such a thing as the "distinctively Christian" and that Christianity is not merely an ongoing multiplicity of interpretations with family resemblances.

There is no simple, unequivocal way to describe Christianity. To be a Christian is not simply to imitate the historical Jesus. Ethical dilemmas cannot be simply solved by a biblicist[1] approach to the Bible. Nor is the Christian religion rightly understood by translating the Word of God addressed to man into a scientific theology. We cannot grasp the essence of Christianity by rendering it into a rationally articulated statement. Prior to the articulation of the faith is the experience of Christ. Nor, again, is Christianity simple pietism, for the individual experience of Christ receives its fullest expression in communion with others.

Schweitzer is to be commended for his attempt both to be true to the historical Jesus and at the same time to link him with the present experience of the Christian. He saw clearly that the Liberals had domesticated Jesus by their benign portrait of him. Schweitzer's portrait of him was anything but domesticated. Yet even Schweitzer's Jesus -- who was a stranger to our time -- still has authority over us.

> Anyone who ventures to look at the historical Jesus straight in the face and to listen to what he may have to teach him in His powerful sayings, soon ceases to ask what this strange seeming Jesus can still be to him. He learns to know Him as One who claims authority over him.[2]

How can the Christian, gazing upon the strange, enigmatic historical Jesus, find in him the source of authority? Schweitzer answers that although the eschatological Jesus must return to his own time, his personality continues to exercise authority through the centuries.[3] Schweitzer also speaks of the spirit of Jesus bridging the yawning chasm between his time and ours.[4] The Christian must surrender to the *will* of Jesus.

> The true understanding of Jesus is the understanding of will acting on will. The true relation to Him is to be taken possession of by Him, Christian piety of any and every sort is valuable only so far as it means the surrender of our will to His.[5]

Bultmann is to be commended for his refusal to subsume Christ's significance for us under ethical regulations. But this very strength of Bultmann also points to a weakness in his general position: its lack of concreteness. For Bultmann, for example, *what* one has to do to love one's neighbour cannot be imposed from without but arises from the situation itself. The person who truly experiences Christ will know what to do. "Truly when a man asks after the way of life there is nothing in particular to say to him. He is to do what is right, what everyone knows."[6] Now it is true that in obvious cases the course of action a Christian should follow is self-evident:

> Whoever sees a wounded man lying on the road knows without further command that it is right to help him. Whoever encounters the sick and oppressed knows that no Sabbath ordinance can hinder the duty to help.[7]

But we live in an age when all choices are not quite so simple. What attitude is the Christian to adopt to giant corporations? What stance should he or she take to nuclear power or genetic engineering? It is certainly not self-evident what a distinctively Christian attitude to these issues should be. The truth of the matter is, of course, that existential Christianity is essentially individualistic and does not speak to humans as members of a complex society.

In attempting to understand the nature of Christianity, we must begin from the givenness of God and the corporate environment of the Church. Now although God may have given of himself decisively in the Christ-event, this does not exhaust his givenness. The incarnation is the givenness of God in a human setting at a particular point in time, but this must not be used as a shortcut to theologizing. To say that God has given, and continues to give, of himself in history to fallible men precludes a docetic Christ, an infallible Church, and a Bible "dropped directly and unrelatedly from heaven."[8] It also precludes two extreme views of the Christian revelation: the view that it is time-bound and immutable (as in the Judaizing heresy) and the view that it is suprahistorical (as in Gnosticism). The former we may label the "classicist" view, and the latter, the "ahistorical."

We argued in chapter two for a dynamic unity of Christianity. Christianity is constantly changing and adapting to new situations and new horizons. The classicist view does not do justice to this dynamism. The ahistorical view, on the other hand, does not do justice to the fact that Christianity is anchored in an historical revelation. Christianity is, as Turner put it so simply, an interaction between fixed and flexible elements. The distinctively Christian must be tested by its coherence with other Christian truths and its correspondence with the faith data.

Early Christianity was not based on a supranaturalist ethic. J. A. T. Robinson calls the supranaturalist ethic "heteronomous," in the sense that "it derives its norm from 'out there.' . . . It stands for 'absolute,' 'objective' moral values."[9] Christianity was from the beginning not only a religious but also a social reality. It was grasped within the concrete, dynamic, and complex reality that is the community of believers. The process of normative self-definition was an ongoing process, subject to adaptation and acclimatization. Its major referent point was Jesus the Lord. As Hans Kueng says:

> What is specifically Christian therefore is the fact that all ethical requirements are understood in the light of the rule of the crucified Jesus Christ. It is then not a question of merely what is moral. The gift and the task coincide under the rule of Jesus Christ; the indicative already contains the imperative. Jesus, to whom we *are* subordinated once and for all in baptism by faith, *must* remain Lord over us.[10]

It is adherence to Jesus that gives Christianity its concreteness and distinguishes it from theosophies and philosophies. Jesus is not a thinker; he does not offer reasoned, rational arguments which stand or fall on their own merits. His message is intrinsically bound up with his person; he solicits belief in himself. He is, as Kueng aptly puts it, "the living archetypal embodiment of his cause."[11]

To be a disciple of Jesus involves following -- not imitating but following[12] -- him, tuning our lives with his. It involves a basic reorientation to life, rooted in thanksgiving for the boon of salvation (which is expressed in the *lex orandi*). Jesus appeals to the whole person -- to the passions, to imagination, to spontaneity. And this is his essential mystery -- Jesus as a concrete person elicits a response which a principle, symbol, or philosophical abstraction cannot, and yet as Lord he remains elusive, never fully grasped. For we must beware of reducing the cause of Jesus to the cause of the Lord Jesus. Schweitzer was undoubtedly right: Jesus expected the *eschaton*, and it did not come. But to say that Jesus was in error is not to say God was in error. In the words of C. K. Barrett:

> It is natural that men should wish to see without delay the happy ending of the story in which they are involved; natural also that those who experienced events so profoundly disturbing as the death and resurrection of Jesus should suppose that any subsequent extension of history could be nothing but an anti-climax. But may it not have been God's intention, unforeseen in its fullest extent even by Jesus himself, that life between the incarnation and the *parousia* should be the normal state of mankind? In this way men, caught in the tension of deferred hope but living by the Holy Spirit, might learn the discipline of the life of the children of God, each generation wrestling afresh with fear and defeat, existing under the shadow of death, and discovering what it is to live by faith. How could the discovery be made in any other setting? *Vivendo, immo moriendo et damnando fit theologus, non intelligendo, legendo, aut speculando. Fit Christianus,* Luther might have written.[13]

PART II

THE PHENOMENON OF MODERNITY

Chapter Three

The Origin: Changing Horizons

In the previous chapters we have seen how in order to relate Christ to the modern world we have to reckon with the change in horizon between the modern world and that of the New Testament. The question of how different the horizon of modernity is from those which preceeded it is one which is of interest not only to the biblical scholar but also to any student of the history of the Western tradition. There are those who argue that, in fact, there has been a fracture in the Western tradition -- a radical discontinuity -- which has implications not only for religious thought but also for political, philosophical, social, and scientific thought as well. C. S. Lewis, for example, in his delightfully lucid inaugural lecture at Cambridge in 1954,[1] specifies four aspects of the modern world which differentiate it from what has gone before. He believes that the understanding of politics is different in the modern era and that this is exemplified in the vocabulary used nowadays. We no longer have "rulers" -- we have "leaders"; we require of them not justice, wisdom, and clemency, but dash, initiative, and magnetism. He also sees significant differences in the arts, inasmuch as no previous age has ever produced such "shattering and bewildering" new art, like that of the Cubists, the Dadaists, the Surrealists, Picasso, and the modern poets. Thirdly, he sees a great religious change, which he calls "the unchristening." We are living in a post-Christian age and are thus cut off both culturally and religiously from the Christian past. Finally, Lewis sees the coming of the machine as heralding a change as momentous as that from the Stone Age to the Bronze.

Arthur Koestler has argued similarly to Lewis, although from a different perspective. Koestler claims that we are on the "hinge of history." Through the dramatic increase in our ability to utilize power, mass communications, and specialized knowledge, plus the upsetting of the ecological balance, we are now faced with a unique situation. It could lead to the death of *homo-sapiens* as a species.[2]

The question of whether there is discontinuity in the Western tradition is an important one. We now turn, therefore, to the changes which have taken place in giving rise to the horizon of modernity, to see if they are as dramatic as some writers suggest.

The last five hundred years have seen changes of all kinds, of course, and it would be foolhardy to try to detail the nature and significance of all of them. But there are various obvious landmarks in the changing horizons of the Western world which have contributed to our contemporary self-understanding. It is these more obvious landmarks that we will examine.

The Scientific Revolution (c. 1540-1700) was, for example, an intellectual revolution of far-reaching consequences.[3] Before it, men had certain established ideas about God, humanity, and the world. After it, many of these ideas were no longer tenable. During the High Middle Ages, knowledge and science had relied very heavily upon the ideas of Aristotle, mostly because of the use made of his ideas by Aquinas. As is well known, Aquinas transposed Aristotle's idea of a First Cause to a study of God as Prime Mover. He used Aristotle's ideas on the natural order to show how the world was ordered according to God's law. This defence of the order of the world was also a useful argument in favour of the *status quo*. At the apex of the hierarchical system of the Middle Ages were the Pope and the Holy Roman Emperor, and by arguing that such a structure was divinely sanctioned, Aquinas was also conserving the power of the Church. But as Arthur Koestler and others have observed, this marriage of Aristotelian physics with Aquinas's theology proved disastrous.[4]

It was in the heavens (the lunar world) that Aristotle saw the supreme example of perfection and orderliness. The earth was the centre of the universe. As the earth was the centre of the universe, so *homo-sapiens* was the centre of creation itself. The human alone had a rational soul and was able to reflect upon purpose and creation. In "Christianizing" Aristotle, Aquinas took over many of his assumptions, including his scientific ones.[5]

The Scientific Revolution changed these views. The revolution is often said to have begun with Copernicus, who in 1543 suggested that the earth was not the centre of the universe and that day and night were caused by the earth rotating on its axis. Copernicus's ideas were attacked by the Church, and even Luther condemned his views as "the over-witty notion of a Fool who would fain turn topsy-turvy the whole Art of Astronomy."[6] Even so, Copernicus's work signalled the beginning of a whole new way of looking at things. No longer was something accepted because it was asserted by tradition and reinforced by the authority of the Church. The new science was based on experiments, observations, and mathematical analyses. This scientific method was given concrete expression in the work of Francis Bacon. And it was Bacon who epitomized the changed attitude towards nature when he summoned us to control nature "for the relief of man's estate."[7] Whereas ancients such as Aristotle had conceived of science as the understanding of nature, Bacon conceived of science as control over nature. It is precisely this view which has been given concrete expression in the technological society. Bacon's view of nature, then, signalled a change of momentous significance.

The avalanche of new ideas, once it began, could not be stopped, even though the Church often tried to suppress scientific learning (as in the famous example of Galileo). Aristotle's emphasis on everything having a First Cause came to be replaced by a mechanistic understanding of the world -- a world which operated on its own, according to certain laws, and which did not need divine assistance or intervention to keep it going. Intellectual and mechanical tools were invented which enabled more exact measurements to be made and which furthered the mechanistic understanding of the world. Such inventions were analytical geometry, calculus, the microscope, the telescope, the thermometer, and the pendulum clock.

It was above all in the work of Isaac Newton that the spirit and achievement of the Scientific Revolution received its fullest expression. Although Newton himself remained deeply religious, his work demonstrated that the world could be understood in mechanical terms. It was a self-regulating mechanism that had no need of a First Cause. The work of Aquinas, which had so laboriously linked a First Cause to the existence of God, became less compelling. When questioned, the tradition of the Church was found suspect. The authority of the Church in scientific matters was successfully challenged. Science was severed from theology, and the process of secularization -- although not fully complete until modern times -- had begun.

The Scientific Revolution was a *Zeitgeist* animated by many people, but following in the wake of the Scientific Revolution were certain outstanding men whose work, brilliant in its individualism and shocking in its implications, proved to be of the greatest significance for modernity. Three such men are Darwin, Freud, and Marx.

1. The Horizon of Suspicion

Charles Darwin's theory of evolution is well known today, as are the controversies surrounding its introduction into Victorian society. And yet there is still an astonishing amount of misinformation perpetrated about what Darwin actually said. The theory of evolution was not original to Darwin (he himself credited Aristotle with having first "shadowed it forth"),[8] and Darwin did not, of course, suggest that man was descended from the ape. It is unfortunate indeed that Darwin should have been so misrepresented and especially unfortunate that his theory should be used for purposes which were not at all in accord with his original intention.[9] The full significance of what Darwin did say, however, can hardly be underestimated. As Langdon Gilkey says:

> The influence of Charles Darwin in creating the sense that we are brought into being by "chance" interactions of blind events is undeniable. More than any other result of modern inquiry, his theory of origins seemed to displace man from his former setting within an eternal rational order, or a purposefully willed law of selection combined with random mutations. The modern sense of radical contingency, relativity, and temporality . . . has its origin in Darwin. Needless to say, back of Darwin lies the entire development of post-Galilean science and philosophy, with its denial that teleology or purpose has any creative role in the understanding of the habits of nature. What Darwin did was to enlarge this anti-teleological view of nature to cover the question of the origin of man's form and thus to include all of our relevant environment, and ourselves, within the blind mechanism of matter.[10]

The true precursors of Darwin can be said to be Charles Lyell and Richard Chambers. Lyell's *Principles of Geology* was written specifically to proffer an alternate theory to Baron Cuvier's theory of catastrophism. Cuvier believed the world to be very young but shaped by periods of catastrophic change. This explained the existence of fossils of now extinct species. While it would be strictly incorrect to describe Lyell's book as "evolutionist," it certainly implied evolution, for he maintained that the world had not been subject to epochal catastrophes, and therefore change of form must have taken place.[11] Chambers's *The Vestiges of the Natural History of Creation* was more explicit:

> The idea which I form [of the progress of organic life] is, that the simplest and most primitive types, under a law to which that of like production is subordinate, gave birth to the type superior to it . . . that this again produced the next higher, and so on to the highest, the advance being in all cases small.[12]

There were two major differences between Darwin's work and that of Chambers. In the first place, Darwin's work was one of great scientific merit; Chambers's was not. Second, Chambers saw providence at work in evolution. While Darwin did not deny the idea of a creator God, he did reject the idea of providence: "There seems to be no more design in the variability of organic beings, and in the action of natural selection, than in the course with which the wind blows."[13]

Most of Darwin's ideas on evolution were formulated while he was an unpaid naturalist on the *Beagle*, a ship which undertook a five-year mission to explore the world. During this voyage, it struck Darwin quite forcefully how species competed against each other for such life-resources as air, sunlight, food, water, and shelter. He further observed how some species were more plentiful than others and that these were almost always better adapted to their environment. After he had experienced an earthquake off the coast of Chile in 1835, the theory of evolution began to crystallize in Darwin's mind, for now he realized not only that species struggle for survival and change while doing so but that environments change also.

Darwin's theories were strikingly confirmed by his visit to the Galapagos islands. There he observed concretely how some species had adapted to their environment. The finches on one island had thicker beaks than those on another. The birds with thicker beaks were found on islands where the food supply was mainly seeds and nuts. The birds with smaller beaks were found on islands where insects were the staple food. These thinner beaks were, of course, much more suitable for rooting out insects. Obviously, then, the birds which were better adapted to their environment were the ones which survived.

It was after he had read Thomas Malthus's *Essay on Population* that Darwin finally knitted together his observations into a theory and was able to think of evolution as the change species undergo through natural selection.

> Being well prepared to appreciate the struggle for existence which everywhere goes on from long-continued observation of the habits of animals and plants, it at once struck me that under these circumstances favourable variations would tend to be preserved, and unfavourable ones to be destroyed. The result of this would be the formulation of new species.[14]

Darwin might well never have published his ideas had he not received from fellow biologist Alfred Wallace an essay which proposed a theory remarkably similar to his own. Upon the advice of his friends, Darwin allowed a summary of his own views to be published along with Wallace's essay in the *Linnaean Society Journal.* He then set to work feverishly and one year later, in 1859, produced his *On the Origin of Species.*[15]

The book created a storm of controversy. Although in the *Origin* Darwin had not specifically focused on human beings,[16] some readers were quick to see that it had dramatic implications for the Christian view of humankind. It was seen to be in direct conflict with the Genesis account of creation. This was most apparent at the meeting of the British Association in Oxford in 1860. Darwin himself was absent from the meeting because of ill health, and his position was defended by Thomas Huxley and Joseph Hooker. The attack on Darwin's work was led by the Bishop of Oxford, Samuel ("Soapy Sam") Wilberforce. The substance of Wilberforce's attack may be gathered from the following words of a review he wrote of *Origin*:

> Mr. Darwin writes as a Christian, and we doubt not that he is one. . . . We therefore pray him to consider well the grounds on which we brand his speculations. . . . First, then, he . . . declares that he applies this scheme of natural selection to man himself, as well as to the animals around him. . . . Now we must say at once, and openly, that such a notion is absolutely incompatible, not only with . . . the word of God on that subject . . . but . . . with the . . . moral and spiritual condition of man. . . . Man's derived supremacy over the earth; man's gift of reason; man's free will and responsibility; man's fall and man's redemption . . . are all equally and utterly irreconcilable with the degrading notion of the brute origin of him who was created in the image of God, and redeemed by the Eternal Son.[17]

Here we see clearly the fears engendered among some clergy[18] by Darwin's theory. If creation (and therefore humanity) was not immutable, then this not only contradicted the account of creation given in Genesis but also seemed inconsistent with the idea of humanity as the apogee of creation, created in the image of God.

Darwin had not, as we have said, dealt directly with humanity in the *Origin*. It was many years later, in 1871, that Darwin's views on humanity were made explicit in his book *The Descent of Man*.[19] In this book Darwin traced the ancestry of *homo-sapiens* back to a common ancestry with the beasts and gave humanity a "pedigree of prodigious length, but not, it may be said, of noble quality."[20] With this explicit formulation of the evolution from lower forms, the Christian view of humanity's fall and redemption was definitely seen to be under attack. As Livingston has commented:

> Darwin's interpretation of nature was infinitely more damaging to the Christian view of the world than the revolutions of either Copernicus or Newton. . . . Darwinism challenged the entire biblical account of man's unique creation, fall, and need for redemption. The doctrine that man was the product of a long evolutionary process from lower to higher species was simply incompatible with the traditional interpretation of the Fall in Genesis. In Darwin's opinion man had risen from a species of dumb animal, not fallen from a state of angelic

> perfection. How could one impute a sinful fall to a creature so superior to his brutish ancestors in intellect and morals? And if man is the *chance* product of natural variation, what sense does it make to say that man is the crown of God's plan, created in the very image of his creator?[21]

Thus, although Darwin's work was not itself either philosophical or ethical, it was perceived as having important philosophical and ethical implications.

Darwin's work had a great impact upon our thinking about ourselves and the world, but the impact of Sigmund Freud's work was even greater. Freud altered our thinking about ourselves in the most radical way. It is often difficult to appreciate the full extent of Freud's influence, but when we pause to think how often we use Freudian terms in everyday speech, we do get some inkling. How often do we hear "wishful thinking," "repression," the "unconscious!" If we now accept that dreams and slips of the tongue and pen have meaning, it is because of Freud. And if we are more open in our attitudes to sex and to the mentally ill, it is in no small measure due to Freud. Freud's life and work have indeed changed the course of Western intellectual history.[22]

In what ways did Freud change the horizon of the Western world? In the first place, he changed dramatically our way of thinking about people who are mentally ill. Indeed, it is true to say that before Freud those who behaved in a neurotic or irrational way were not treated as people at all -- they were often placed in asylums for the insane and forcibly restrained. Their strange behaviour was often seen as resulting from physical causes or, in some cases, demonic possession. Freud perceived the mentally ill as sick *people*; in his method of treating them by non-physical means -- psychoanalysis -- he was to plumb the depths of human behaviour in such a profound manner that we can never again reflect upon ourselves in the same way.

Freud made the most significant contribution to the understanding of the human animal with his theory of the unconscious. He never claimed to have discovered the unconscious -- the poets and philosophers had long known of it -- but he did suggest phenomena which the unconscious helped explain. For example, many memories were not accessible to ordinary introspection but could be evoked by certain techniques, such as free association or hypnotism. And there was, of course, the evidence from what he called the psychopathology of everyday life, e.g., slips of the pen or tongue, lapses of memory, etc.

The significant thing about Freud's theory of the unconscious was that it helped explain the behaviour of all people -- not just neurotics.

Upon reflection, we have all encountered (both within ourselves and in conversation with others) intense resistance to certain desires or memories. Certain topics of conversation will evoke annoyance, inattention, feigned fatigue, or, very often, rationalization. That is, a thought or action will be justified by an appeal to reason or practical necessity, when it is, in fact, motivated by irrational desire. It was this process that Freud uncovered and in doing so opened Pandora's box, for he lay bare the workings of the human psyche in a most uncompromising way. Humans were revealed to be not so much creatures "noble in reason," but rather creatures driven by irrational forces often beyond their grasp and control.

It was the analysis of dreams that provided Freud with the clues which led to his theory of the unconscious. Freud, in fact, described the interpretation of dreams as "the royal road to the unconscious." In his famous book, *The Interpretation of Dreams,*[23] Freud demonstrated two important things. First, dreams have a purpose or meaning. Second, the ***manifest content*** of the dream is a disguised expression of the *latent content* (i.e., unconscious wishes or instinctual desires). The mechanisms whereby the latent content of the dream is transposed into the manifest content are now well known. Often, for example, abstract ideas are ***dramatized*** and transformed into concrete images. Many elements in dreams are ***symbols***. And often there is ***condensation*** or ***displacement***.

Freud's original model of the mind consisted of two levels: the conscious and the unconscious. He had thought that the repressing forces were part of the conscious mind. But he discovered that patients could not, with an effort of will, bring buried memories to the conscious mind. Moreover, often some things which had been repressed and finally brought into the conscious mind were not shocking to it. The "censor" was apparently less enlightened than the conscious mind! Freud therefore substituted for his earlier twofold division a threefold division of the psyche into the *Ego*, *Superego*, and the *Id*.

The Id is the inherited instinctual part of us. It consists of basic appetites and desires and is almost wholly unconscious. It wants only satisfaction and is amoral and illogical. Towards the end of his life Freud saw human life as a continual struggle between the life (*Eros*) instinct and the death (*Arakne*) instinct.

The only part of the mind of which we have much knowledge is the Ego. This is conscious, civilized, and rational. It thus seeks to gain control over the Id, which it recognizes as potentially destructive to the human organism.

The Superego is a kind of "primitive conscience." It is almost wholly unconscious and non-rational. It arises out of the demands of others, especially the parent of the same sex. The kind of authority imposed upon the child by this parent characterizes the Superego. The parent who has been overdemanding of the child will create an overdemanding Superego. In later life a Superego with harsh and irrational standards may clash with the more rational Ego. Sometimes this has no damaging consequences (e.g., the engendering of strange pangs of guilt for working on Sundays or lying in bed on Saturday morning), but at other times the clash may lead to neurosis.

The Id, the Ego, and the Superego are concepts that are well known in the modern world, as is Freud's most controversial psychoanalytical concept, the Oedipus Complex. The child has an ambivalent attitude towards its parents. At first, the totally dependent infant is strongly attached to the mother. A little later the male child develops a strong sexual attachment for the mother. The child sees the father as a rival -- a strong and powerful rival. He hates and fears his father, and only later can he overcome this hatred and incestuous relationship with his mother through identification with the father. Freud believed that at the core of all mental illness was an unresolved Oedipus Complex.

Despite the fact that the universality of the Oedipus Complex has been disputed by almost all reputable scholars,[24] Freud remained convinced that it was the crucial key not only to mental illness but also to the genesis of religion. Freud's views on religion are particularly interesting for our inquiry, and it is to those views that we now turn.

Freud's earliest views on religion were expressed in a paper entitled "Obsessive Acts and Religious Practices,"[25] in which he pointed out the similarities between religious ritual and the neurotic rituals arising out of obsessional neurosis. Although there are obvious differences between the two rituals, notes Freud, there are also very evident similarities in that any deviation from the strict pattern established in the rituals results in anxiety or qualms of conscience. Why? Religious ritual has meaning -- explained by the clergy -- and thus deviation from the set pattern violates the significance of the act. This results in qualms of conscience. Freud's great insight into neurotic ritual was to see that it, too, had meaning. A repressed instinctual sex impulse will manifest itself in neurotic ritual. The neurotic ritual is, in fact, a manifestation of a compromise between the warring factions of the mind: the impulse is partially blocked and partially gratified. Freud believed that this was true of religious ritual also; that is, through the ritual there is displacement and attempted renunciation of socially harmful instinctual impulses. Concludes Freud:

> In view of these similarities and analogies one might venture to regard obsessional neurosis as a pathological counterpart of the formation of a religion, and to describe that neurosis as an individual religiosity and religion as a universal obsessional neurosis.[26]

Having settled early on the idea that religion was an obsessional neurosis, Freud somewhat later in his *Totem and Taboo*[27] attempted to expose the origin of religion. In the most primitive form of religion -- Totemism -- the man is bound to take a wife outside his own clan (exogamy), and the eating of the totem animal is prohibited. These taboos are, however, often suspended at the annual religious celebration. Through the latent content of the ritual, Freud sees the origin of religion in the Oedipus Complex. In fact, Freud sees the beginnings of all religion, ethics, and society in a traumatic event which occurred at the dawn of time.

At the dawn of time, says Freud, humans lived in groups similar to the gorilla horde -- the strongest male dominated a group of women and drove away all rivals, including his sons. One day the expelled sons banded together and killed the dominant father. They appropriated his power by eating his body and taking his women. But then they felt remorse and guilt, and so "they undid their deed by declaring that the killing of the father substitute, the totem, was not allowed, and renounced the fruits of their deed by denying themselves the liberated women."[28] Although the forces of repression have hidden this event in the human psyche, psychoanalysis is able to root it out and expose the great murder which gave rise to all religious life.

This theory of Freud's has received scant support.[29] The attack upon religion in *Totem and Taboo* is certainly not to be taken as seriously as his other major attack in *the Future of an Illusion.*[30] Here, arguing that the human's instinctual wishes include incest, cannibalism, and lust for killing, Freud asserts that the renunciation of these instincts requires recompense and that the function of religion is exactly that -- it seeks to compensate man for renunciations demanded by civilization. By projecting a father-figure (God), religion alleviates the terrors of nature and fate and also promises recompense in the form of an afterlife. Although religion performed a useful function for humankind in its early stages of development, Freud declares (in a true Enlightenment manner) that the rule of reason must now displace the rule of religion. Religion is the universal obsessional neurosis of humanity[31] which it must overcome if it is to attain to maturity.

A summary of Freud's criticism of religion is found in his *New Introductory Lectures on Psycho-Analysis*:

> While the different religions wrangle with one another as to which of them is in possession of the truth, in our view the truth of religion may be altogether disregarded. Religion is an attempt to get control over the sensory world, in which we are placed, by means of the wish-world, which we have developed inside us as a result of biological and psychological necessities. But it cannot achieve its end. Its doctrines carry with them the stamp of the times in which they originated, the ignorant childhood days of the human race. Its consolations deserve no trust. Experience teaches us that the world is not a nursery. The ethical commands, to which religion seeks to lend its weight, require some other foundation instead, for human society cannot do without them, and it is dangerous to link up obedience to them with religious beliefs. If one attempts to assign to religion its place in man's evolution, it seems not so much to be a lasting acquisition as a parallel to the neurosis which the civilized individual must pass through on his way from childhood to maturity.[32]

Freud's views on religion are open to all kinds of criticism. The Primal Horde Theory will simply not stand the test of historical scrutiny. Nor will his Oedipus Complex. Empirical religion is a great cross-cultural mixture, something which is certainly not apparent from Freud's writings. In particular, the idea that, if we can bring out the causes of a man's beliefs, we have then discredited the reasons for them is a serious fallacy. To quote Alasdair MacIntyre:

> To classify a belief as wish-fulfilment is not to speak of its truth or falsity, but is to say that it is held from certain motives which have nothing to do with its truth or falsity. And thus Freud's explanations of the origin of religious belief -- or any other explanation if Freud is finally discredited -- throws light on the irrationality of religious believers rather than on the untruth of religious beliefs.[33]

Even though Freud's criticisms of religion turn out to be somewhat ill-founded, his general impact upon Western civilization has been considerable. Yet even Freud's impact cannot be said to have been greater that that of Karl Marx. It was, after all, Marx's work which inspired the great political revolutions at the beginning of the century and which ultimately led to the political polarizations of the modern world. The history of the modern era would be inexplicable without some understanding of Marx's part in it. Much has, of course, been written about Marx, and it is not our intention to deal with all aspects of his thought here. But what Marx said about religion in fact illuminates his thought as a whole, for his critique of religion is the model for his critique of secular illusions.

In his doctoral dissertation Marx says:

> Philosophy does not make a secret of it. The profession of Prometheus, "In simple words, I hate the pack of gods," is its its own profession, its own aphorism against all divine and earthly gods who do not acknowledge human self-consciousness as the highest divinity. It allows no rivals.[34]

For Marx, the Middle Ages was "the period of consummate unreason."[35] Christianity was irrational; it was, indeed, the very expression of irrationality. This is why in the above quotation Marx vehemently denounces any attempt to reconcile philosophy with theology as an insult to reason. Moreover, Marx's attacks upon religion were not mere *ad hoc* affairs; they were quite integral to his whole way of thinking about the world. Hence he says, "For Germany, the criticism of religion has been largely completed, and the criticism of religion is the premise of all criticism."[36] By this Marx means that the criticism of religion is necessary and crucial, since humanity cannot change the world until it rids itself of its illusions about it. It is in this sense that religious critique was of vital importance both in itself and as a model for the critique of secular illusions. Religious consciousness was a false way of looking at the world, and it was important for people to see the world for what it was.

Marx accused his teacher Hegel of laying the philosophical foundations of "false consciousness." Hegel's philosophical system began with thought. The concrete, the human being who exists in the world and with others, was the object of thought. Marx says, when discussing Hegel's great work *The Phenomenology of the Spirit*:

> This process must have a bearer, a subject, but the subject first emerges as a result. This result, the subject knowing itself as absolute self-consciousness, is therefore God, absolute Spirit, the self-knowing and self-manifesting idea. Real man and real nature become mere predicates, symbols of this concealed unreal man and unreal nature. Subject and predicate have, therefore, an inverted relation to each other.[37]

In other words, Hegel made thought into subject and being into a predicate. In Marx's eyes this was to put the cart before the horse -- it is being which precedes thought; it is the human being who thinks. Marx insisted that philosophy must begin with the centrality of the human being, and he therefore stood Hegel "on his head" and reinstated the human being as the subject and thought as the predicate.[38] When this was done it was seen that religion was a result of human creativity. In a philosophical system such as Hegel's, religion could only confront humanity as something independent of its own productivity. When religion was conceived as something not of humanity's own devising, it stood over against humanity as an independent (and alien) force. Humanity became alienated from something which it had in fact created itself. Such a misconception impoverishes humanity.

Marx saw this description of religious alienation as a useful analogy for the worker's alienation. Humanity's impoverishment by an alien being was analogous to the worker's situation in society. The alien being in the worker's situation was the commodity:

> There is a definite social relation between men, that assumes, in their eyes, the fantastic form of a relation between things. In order, therefore, to find an analogy, we must have recourse to the mist-enveloped regions of the religious world. In that world the productions of the human brain appear as independent things endowed with life, and entering into relation both with one another and the human race. So it is in the world of commodities with the products of men's hands. . . . The religious world is but the reflex of the real world. And for a society based upon the production of commodities, in which the producers in general enter into societal relations with one another by treating their products as commodities and values, whereby they reduce their individual private labour to the standard of homogenous human labour -- for such a society, Christianity, with its *cultus* of abstract man, more especially in its bourgeois developments, Protestantism, Deism, etc., is the most fitting form of religion.[39]

Religion is a form of false consciousness; indeed, it is the classic form of false consciousness. But false consciousness is not limited to religious consciousness. Marx elaborated his theory of false consciousness into a theory encompassing all ideologies. An ideology is a set of beliefs that sponsors a particular set of economic arrangements which the dominant classes wish to conserve:

> For instance, in an age and in a country where royal power, aristocracy, and bourgeoisie are contending for mastery and where, therefore, mastery is shared, the doctrine of the separation of powers proves to be the dominant idea and is expressed as an "eternal law."[40]

Although ideologies often purport to express universal truths, they in fact express transient truths. Ideology is false consciousness, a false way of looking at the world. Religious consciousness is false because it is an "inverted world consciousness."[41]

To put it in a nutshell: Marx argued that only the radical critique of society could complete the critique of ideology which includes not only religion but also all forms of secular illusion. Marx saw an atheistic society as both possible and desirable. It is possible because atheists can be respectable men. It is desirable because it is not atheism but religion which debases humanity.[42] But Marx did not wish to make atheism an end in itself; rather it is both a prerequisite for communism and later an essential element of communist society. To quote Marx himself: "Communism begins where atheism begins . . . but atheism is at the outset still far from being communism."[43] This exemplifies Marx's dialectical understanding of history: communism is the negation of capitalism, and atheism is the negation of religion. Both complement each other and represent the next stage of history -- communist society. But neither witnesses the final stage of human development, the classless society:

> Once the essence of man and of nature, man as natural being and nature as human reality, has become evident in the practical life, in sense experience, the quest for an alien being, a being above man and nature (a quest which is an avowal of the unreality of man and nature), becomes impossible in practice. Atheism, as a denial of this unreality, is no longer meaningful, for atheism is the negation of God and seeks to assert by this negation the existence of man. Socialism no longer requires such a roundabout method; it begins from the theoretical and practical sense perception of man and nature as essential beings. It is positive human self-consciousness attained through the negation of religion; just as the real life of man is positive and no longer attained through the negation of private property, through communism.[44]

In other words, atheism -- which asserts the existence of humanity through the negation of God -- vanishes in the classless society because that society is the realization of the human being as a social being.[45]

What is significant about Marx's thought for our purposes is that he not only radically questioned the nature and function of religion in the Western tradition, but also questioned the very basis of social and political structures. Even if one does not ultimately accept Marx's critique as correct, one has to accept that after Marx we can never again think of society in quite the same way. He introduced suspicion into our thinking about the nature of civil society and religion.

In fact, all three of the men we have examined -- Darwin, Freud, and Marx -- have left us a legacy of suspicion. All three men have made us suspicious of traditional wisdom. Traditional wisdom --

through Christianity -- had asserted that we were created in the image of God. Darwin's work led to this being questioned. Traditional wisdom -- following Aristotle -- had asserted that we were creatures "noble in reason." Freud's work led to this also being questioned. "Freudian Man" was hardly a rational creature who could deliberate dispassionately upon himself and his world, but rather a creature driven by irrational desires and impulses. Finally, Marx questioned the traditional way of looking at social and political structures and debunked the idea that in some way they were divinely sanctioned. This legacy of suspicion, bequeathed to us by Darwin, Freud, and Marx, is a key element in the understanding of modernity. It has contributed in no small measure to the abandonment of traditional values and the emergence of new ones which have cast a new light on religious, philosophical, and moral discourse. It is to the new values of modernity that we now turn.

2. *The Horizon of Modernity*

Many people claim that one cannot speak of "modernity" as though it had an essence, for it has no unity. They argue that if there is one characteristic of modernity, it is its plurality.

This is certainly a point of view worthy of consideration, if only because it is such a popular one. But the fact is that the plurality of the modern world is more apparent that real.[46] Modernity does have a unity, and its essence can be discerned by comparing it with premodern (or classical) political philosophy.[47] If classical political philosophy has a unity -- which is demonstrable -- and if modernity was ushered in by those who broke consciously with it, then modernity, too, has a unity, even if only by reflection.

Classical political philosophy was a "quest for the best political order, or the best regime as a regime most conducive to the practice of virtue or how men should live, . . . according to classical political philosophy the establishment of the best political regime depends necessarily on uncontrollable, elusive fortuna or chance."[48] It was these two assumptions of classical political philosophy -- that it should concern itself with how men *ought* to live and that the establishment of the best regime cannot take place without the consent of Fortuna herself -- which were challenged in the most radical way by the great precursor of modernity, Niccolò Machiavelli.[49]

The importance of Machiavelli as a political philosopher is universally recognized. But the exact intention of his teaching is disputed. In former years Machiavelli was seen as a teacher of evil, and, indeed, his name has become synonymous with evil or untrustworthiness. But in more recent years it has become the vogue to see this interpretation as somewhat quaint or old-fashioned. Machiavelli has become rehabilitated to such an extent that a writer such as Whitfield can argue that he was a humane, democratically-minded republican.[50] Yet there should be no equivocation about the teaching of such an important thinker as Machiavelli: what he taught was evil.[51] The refusal of many to recognize this fact is due primarily to two factors. In the first place, Machiavelli is an attractive personality. One cannot read his correspondence without being drawn to the man. But this should not blind us to the content of his teaching. Attractive people often say the most unattractive things. Second, if we fail to see the truly evil dimensions of Machiavelli's teaching it is because we ourselves are so machiavellian. This is an age when the end of politics is to gain power. In pursuit of that goal, politicians tell the people not what they should hear, but rather what they want to hear. In the "business" of getting elected, truth is often detrimental. No wonder Machiavelli's teaching does not shock us. To repudiate Machiavelli would be to repudiate what we are.

What was it that Machiavelli said that was unique? Francis Bacon captured its essence beautifully when he wrote: "We are much beholden to Machiavelli and others, that write what men do, and not what they ought to do."[52] Machiavelli knew exactly what he was saying. His intention is not opaque to the alert reader. His books on political philosophy -- *The Prince* and *The Discourses on the First Decad of Titus Livy* -- are different in many superficial respects but not in substance.[53] The message in both books is clear -- to follow the teaching of classical political philosophy can only lead to ruin:

> A great many men have imagined states and princedoms such as nobody ever saw or knew in the real world, for there's such a difference between the way we really live and the way we ought to live that the man who neglects the real to study the ideal will learn how to accomplish his ruin, not his salvation.[54]

Machiavelli therefore sets himself the task of redefining the horizon within which political life should take place. One should begin with what people are, not with what they ought to be. People are basically depraved.[55] This depravity cannot be cured solely by laws. Laws, to be effective, have to be enforced. It is therefore force and its use which is the key to the exercise of political power. *The Prince* is especially concerned with the art of obtaining power and then using force (and every other unscrupulous means at one's disposal) to establish the regime the Prince desires. Fortuna is a woman who will surrender to the man of resolute will and determination.

No one can deny that political life is, and always has been, full of treachery and crime. But those who thus defend Machiavelli on the grounds that he is merely being honest and "telling it like it is" miss the point. To acknowledge that political life is full of treachery is quite different from teaching the art of political treachery. No one before Machiavelli had undertaken such a task because of the belief that what political life is is not what it ought to be. Machiavelli's insistence that we focus on what humanity is, not what it ought to be, was his greatest single contribution to the history of Western political thought. It follows from such reasoning that there is no natural end of humanity. Humanity determines its own ends. Therein lies Machiavelli's significance for our times, for as Warren Winiarski says:

> As transcendent goals for human life are abandoned . . . human life as such is divinized, made into something transcendent; and it is thus that the sciences and the technological arts receive an imperious ordinance to gratify a proliferation of human "needs."[56]

In Machiavelli the political problem has become a technical problem.[57] The political problem can be resolved by lowering the goal of political life and by conquering chance. The conquest of chance requires that science be deployed for the maximum control of the limitations of human life. With Bacon's call to conquer nature "for the relief of man's estate," science is no longer knowledge for its own sake; rather, the end is power. But power is not the only end. One shows one knows by demonstrating mastery. To give this its most radical formulation: one knows only what one can make. "Knowing *equals* making."[58]

The reduction of the moral and political problem to a technical problem, and the vision of the conquest of nature for the relief of man's estate, was followed by an even more significant development: the assertion that what was fundamentally characteristic of humanity was its *historicity*. Humanity has no unchanging essence; it is subject to the change we see in all of nature. We can therefore make no generalizations about the meaning of human life. To use Nietzsche's phrase, we have to accept the finality of becoming.[59]

It was Nietzsche who saw most clearly the implications of the historicity of humanity. Natural science had shown that one need not presuppose purpose in unravelling the mysteries of non-human nature. But humanity had been reluctant to admit that this applied to it too. Humanity clung to the idea that existence had purpose and that purpose was found in rationality.

Nietzsche's treatment of these issues has profound significance for the modern era. He greatly admired Greek tragedy, which showed optimism for what it was -- self-deception. In Greek tragedy the world is seen for what it is -- chaos, not cosmos. Humanity's lot is to suffer. Yet the Greek tragedies sought to inspire humanity resolutely in the face of this suffering. The Greek tragedies offered humanity a vision of nobility, but Socrates -- the destroyer of Greek tragedy -- offered rationality instead. Nietzsche attacks Socrates because he championed the idea that not only did human existing have purpose, but rationality could reveal that purpose. An even greater calamity than Socrates came in Christianity, which Nietzsche dubs "Platonism for the masses." But two thousand years of Christianity are coming to an end:

> Have you not heard of that madman who lit a lantern in the bright morning hours, ran to the market place, and cried incessantly, "I seek God! I seek God!" As many of those who did not believe in God were standing around just then, he provoked much laughter. . . . The madman jumped into their midst and pierced them with his glances.
>
> "Whither is God" he cried. "I shall tell you. *We have killed him* -- you and I. All of us are his murderers. But how have we done this? How were we able to drink up the sea? Who gave us the sponge to wipe away the entire horizon? What did we do when we unchained this earth from its sun? Whither is it moving now? Whither are we moving now?
>
> . . . Are we not straying as through an infinite nothing? Do we not feel the breath of empty space? . . . God is dead. God remains dead. And we have killed him. . . . There has never been a greater deed; and whoever will be born after us -- for the sake of this deed he will be part of a higher history than all history hitherto."[60]

When Nietzsche asserts that God is dead, he is not speaking simply of the demise of belief in the Christian God. He is asserting that the historic sense has destroyed belief in a transcendent ground of permanence. He is dealing with the question of whether there is a progressive character or rationality in the historic process. He is asseverating that the horizons within which we live are our own creations. Horizons are not discoveries about the nature of things; rather, they express the values which humans will. Human beings as historical creatures cannot know what they are fitted for and thus "make" themselves as they go along.[61] This is the crisis of modernity for Nietzsche. A dreadful darkness envelops humanity once it is realized that its horizons are created by human beings themselves. Humanity's will is blighted, for it lacks controlling purpose.

Two basic types of human beings are left at the end of the era of rationality: the "last men" and the nihilists. The last men are those who cling to the ideas of happiness and equality, despite the destruction of the rational bases for such concepts. Their happiness is of a debased sort, for it is bought at the expense of nobility and greatness. The last man is most despicable because he cannot despise himself:

> "We have invented happiness," say the last men, and they blink. They have left the regions where it was hard to live, for one needs warmth. One still loves one's neighbour and rubs against him, for one needs warmth.
>
> Becoming sick and harbouring suspicion are sinful to them: one proceeds carefully. A fool, whoever still stumbles over stones or human beings! A little poison now and then: that makes for agreeable death.
>
> One still works, for work is a form of entertainment. But one is careful lest the entertainment be too harrowing. One no longer becomes poor or rich: both require too much exertion. Who still wants to rule? Who obey? Both require too much exertion.
>
> No shepherd and one herd! Everybody wants the same, everybody is the same: whoever feels different goes voluntarily into a madhouse. "Formerly, all the world was mad," say the most refined, and they blink.
>
> One is clever and knows everything that has ever happened: so there is no end of derision. One still quarrels, but one is soon reconciled -- else it might spoil the digestion.
>
> One has one's little pleasure for the day, and one's little pleasure for the night: but one has a regard for health.
>
> "We have invented happiness," say the last men, and they blink.[62]

Although the last men (the inheritors of rationalism in its last form) predominate, there are those who know and understand what the end of rationality means. They know there is no ultimate purpose to their willing, but nevertheless they must will. They are the nihilists and are admired by Nietzsche more that the last men. Most assuredly one does not need to be a particularly perspicacious individual to see the nihilists and the last men everywhere in the modern world.

The thought of Nietzsche is immensely rich, and our description of it certainly does not do justice to its complexity.[63] But our purpose is not to summarize the thought of Nietzsche but to see what light he and Machiavelli shed on the contours of modernity. It is evident from what has been said that modernity has a substantial identity of its own. There is a certain understanding of existence, the world and humanity's place in it, which pervades modernity. In the modern world knowledge is power: what *can* be done *is* done. The control of nature through technology has become an end in itself. Ecological problems and energy crises are seen as merely symptomatic of an incomplete stage in technology; the correction of the imbalance merely requires a more comprehensive technology. Inherent in technology itself there is a drive towards totality.[64]

It is equally evident that in the modern world happiness is seen as the goal of existence and that this has nothing to do with virtue. As Nietzsche intimated, it is possible to make human beings happy amid the most appalling deprivations of mind and spirit. There are numerous science fiction works which paint an ironic picture of the ultimate future of the "happiness-oriented" society. Zamyatin's *We* and E. M. Forster's *The Machine Stops* are the classical works of this genre, but the best known is Huxley's *Brave New World.* Huxley offers us this vision: a stable and homogenized society, whose members are blithely happy; they fornicate, "get high" on soma, and have every desire gratified. They do not read, write, think, or love. They are, in fact, dehumanized but unaware of it. The frightening thing about the novel is that many of the techniques Huxley describes are no longer of the distant future.[65]

The people in Huxley's Brave New World are happy slaves. They do not want for food; they do not suffer disease; they are not physically brutalized. They are free from adversity. But they are not free. It is precisely at this point -- on the question of freedom -- that we see the character of modernity most clearly. For there is no doubt that the liberal ideal undergirds modernity. Liberalism is, to use George Grant's definition, "a set of beliefs which proceed from the central assumption that man's essence is his freedom and that therefore what chiefly concerns man in this life is to shape the world as he wants it."[66]

Liberalism is the doctrine of open-ended progress. It simply has no conception of a human nature which *ought* to be realized. It affirms so severely the freedom of humanity to make of itself what it will that it eschews any notion of good which might limit such activity. In the nineteenth century liberalism was the doctrine which freed men to exploit each other economically without governmental restraint; "the passion emancipated was the passion of greed."[67] In the present era, liberalism is the doctrine which frees technology from governmental restraint. Mastery of human and non-human nature is the end; this end is not grounded in any notion of good but in some vague idea of happiness and comfort "for the relief of man's estate." This idea is so vague that it is quite incapable of checking technological progress. "The passion emancipated is the passion to innovate."[68]

The irony is indeed painful, as Leon Kass observes:

> Our conquest of nature has made us the slaves of blind chance. We triumph over nature's unpredictabilities only to subject ourselves to the still greater unpredictability of our capricious wills and fickle opinions. That we have a method is no proof against our madness. Thus, engineering the engineer as well as the engine, we race our train we know not where.[69]

The conquest of chance is seen as the chief means of improving humanity, and yet there is no sense of the direction the "improvement" should take. How could there be, when the very idea of the "distinctively human" seems to have been so successfully debunked? For it is at this very point that we see the full significance of the work of Darwin, Freud, and Marx. Followers of Darwin maintain that Darwin's theory shows that the distinction between humanity and the animals is, at best, rather hazy. Humanity shares the same brute origin as that of all the animals. Followers of Freud maintain that Freud's theories show humanity to be driven by the same primitive desires that motivate all animals. Rationality is merely a superstructure built upon irrational fears and desires. Followers of Marx maintain that previous ideas of the good were historically conditioned. Tracts on the "just society" were often mere spurious apologies for bourgeois morality.

There is nothing new about reductionism or relativism. Such views were known to classical philosophy. But the novel feature in the modern situation is that, as we have seen, science seems to have validated these views. The modern age thus affirms the pursuit of rationalized technique and affirms simultaneously that questions of purpose lie beyond rational discourse. We are the true heirs of Machiavelli: we begin where we are, not where we ought to be. We are the true heirs of Nietzsche: we affirm that it is in the realm of history we become what we truly are. Our freedom lies in making the world as we will.

Chapter Four

The Development: Towards a Technological Future

1. Some Theological Perspectives on Modernity and Autonomy

The argument has thus far been that modernity originated in the break with the classical view of the world. The Scientific Revolution repudiated classical science (epitomized by Aristotle). Darwin, Freud, and Marx repudiated traditional views on creation, humanity, and the origin and purpose of society. The teaching of Machiavelli was a devastating attack upon classical political philosophy. And in Nietzsche we see the finest expression of one interpretation of modernity: once we accept the full implications of our own historicity, we must accept the task of creating our own horizons of meaning.

There are those who argue, however, that modernity actually originated in the theological break with tradition epitomized in the theology of such as John Calvin and Martin Luther. In his famous essay, *The Protestant Ethic and the Spirit of Capitalism,*[1] Max Weber argued that the emergence of the modern capitalist spirit could be traced back to the theology of Calvin or, more precisely, to a distortion of Calvinist theology. Some have further maintained that both Calvin and Luther signalled the decisive break with the past and paved the way for modernity by their rejection of natural law. Others claim that modernity originated in Luther's doctrine of the two realms (which denied ecclesiastical authority over secular matters) and his doctrine of a "calling" which made "the world" a place of religious vocation and encouraged participation in worldly affairs.

To find the roots of modernity solely in the Reformation is undoubtedly to overstate the case.[2] Yet Luther's thought in particular does have implications for the debate on the relation of Christianity to modernity. These implications are seen especially clearly in the work of Friedrich Gogarten, one of the most original interpreters of Luther's thought in contemporary theology.

Gogarten insists that there must be no attack upon technology in the name of Christianity, but he never tires of pointing to the dangers of what he calls "secularism" (which he carefully distinguishes from secularization). His starting point is the falsity of what he terms "subjectivism." Christian faith is not grounded in "objectivity" or historical "reality." In true Lutheran style Gogarten characterizes faith as the personal knowledge of the Christ, the one who justifies. "Subjectivism" derives from Descartes's subject/object dichotomy. Through Descartes's influence, we have come to think of the world as "our" object. This leads to a world-view in which we see ourselves as independently reigning over the world because we will to do so, not because we receive it as an inheritance.[3] We are a unity of receptivity and activity. As active agents, we are fully autonomous; as receptive beings we are fully defined by openness -- openness to others and to the mystery of being in the world.[4] When receptivity is eclipsed by activity, one has "subjectivism." Larry Shiner explains:

> Gogarten's point is simply that if God is in fact the creator, then we stand even now in his power. Since self-grounded freedom and autonomy are the unquestioned presuppositions of modernity, theology will serve the Christian community by reminding it that God deals with man precisely in this experience of freedom and autonomy and not in some kind of religious or supernatural realm. So long as Christians are alive to the historicity of God they will not take the message of the New Testament as a substitute for autonomous responsibility, but as its fulfillment.[5]

The "historicity" of God is in fact quite central to Gogarten. The Christian faith is to be understood historically, not metaphysically. The relation of God to humanity and the world is to be defined historically: "to think historically is the effect of the Christian faith on human existence."[6] We exist historically only when we are *responsible for our own destiny.* Hence the significance of the creation accounts in Genesis -- they "depotentiate" the cosmic forces and liberate us from their encompassing power. In the Genesis accounts of creation, it is maintained, God creates the world and gives us stewardship over it. Thus nature is not a sphere populated by divine powers, but a creation of God. The full significance of this view of nature is not fully grasped, however, without the work of Christ. Faith in Christ frees us from bondage to the world and frees us for responsibility for it. Gogarten's favourite text is Gal 4:1-5, which he interprets to mean that it is through Christ that we are delivered from bondage and freed for responsibility; we are heirs who receive the world as an inheritance.

It is at this point that we arrive at a most crucial aspect of Gogarten's thought. Somewhat similarly to Bultmann, Gogarten sees the "documentary" or antiquarian approach to history as inadequate, especially for the proper understanding of Christian faith. Christian faith is grounded in the event of Christ, but this event is not "objective" to us; it is not something we can appropriate by simple historical construction. The event of Christ is present to us. It is present through justifying faith. Without justifying faith we are subject to many "gods" and many "lords" (I Cor 8:5). Furthermore -- and this is most significant -- without justifying faith we seek to win salvation by appropriate deeds. Gogarten sees secularization as a thoroughly authentic Christian development in that it frees us from "works of the law" invested with religious meaning. Gogarten's interpretation of Paul -- and especially of Galatians -- is that the world is handed over to us and that we preside over it through reason. The gospel attributes no saving significance to works, and thus Christian faith secularizes the works of the world and hands them over to human reason. In this way we are freed for God: "The freedom of the son for God and the freedom of the heir toward the world."[7] The liberation of Christians comes through the realization that works in the world do not earn salvation. Paul Schilling comments:

> In faith he [man] now sees both himself and the world as rooted in the creatorhood of God. Hence the world no longer stands over against man as a final reality, or as an eternally valid order which now incloses him and to which he is responsible. Rather it is the creation of God *for* which he is under God responsible, and *in* which he is called to live as God's mature son. In it, therefore, he is not subject to the demands of an external law, dependent on his own works for salvation, but free for the service of God. This is true "independence toward the world."[8]

Gogarten's affirmation that at the heart of the gospel message is salvation through faith is typically Lutheran. His distinctive contribution lies in his argument that secularization should be embraced by the Christian, for it sharply highlights the distinction between faith and works. Moreover, Gogarten's distinction between secularization and secularism is vital. Secularism constructs a world-view which claims to encompass all reality. In secularism reason is seen as able to answer all questions about reality, and this leads to reliance on our achievements for salvation -- the very thing that the distinction between gospel and law denies.

The true meaning of faith lies in the exposure to an open future. Indeed, it is only when we expose ourselves to the mystery of the impenetrable future that we expose ourselves to God or, more exactly, "the on-coming futurity of God."[9] "Futurity is God's nature."[10] This particular characterization of faith is surely worth emphasis, for, as Schilling comments, "Faith is really surrendered if it is changed into knowledge of ethical truths grounded in ultimate reality, hence robbed of its questioning uncertainty and its unconditional openness to the future."[11]

Gogarten contends that there can be no return to the idea of a supernaturally ordered cosmos. The world is profane and must not be invested with religious significance. The horrors and terrors of our modern era have come about through the resacralizing of political and social institutions. Differentiating faith from works thus becomes all the more urgent. Gogarten's insistence on this point, and his interpretation of it in the context of modernity, is a great merit of his work.[12] He has, moreover, attempted to show that our autonomy in the world -- correctly understood -- is not anti-Christian but is, in fact, an integral part of the gospel.

Gogarten's work has not been widely read outside the privileged circle of professional theologians. His ideas have, however, become widely known through the more popular writings of Harvey Cox. Cox openly acknowledges his debt to Gogarten, and he has rendered a great service by giving concrete expression to Gogarten's thought through illustrations from the urban and political life of North America, especially in his well-known book *The Secular City.*[13] Cox presented his basic thesis in terms which were easily understood. The modern secular spirit is, in fact, a development from authentic Christian faith. "Secularity"[14] should be embraced by the Christian. Properly understood, biblical faith desacralizes nature and affirms our historicity. The realm of nature is thus stripped of the mythically derived reverence it received in ancient religion and becomes a creation of God which is put at our disposal. Authentic Christianity frees us for "creative responsibility" in regard not only to our own historicity but also to our use of nature. This aspect of Christianity was obscured during the medieval period, during which time Aristotelian metaphysics gained ascendance, but it has re-emerged into the clear light of day with the rise of secularity. Cox argues that biblical theology "disenchants" the world of nature, "desacralizes" politics, and "deconsecrates" values, leaving life pragmatic and profane. The Christian has no need to fear the technological society. As Cox writes in one of his articles:

> It has been contended by some that a technological civilization does not permit the possibility of Christian faith, that it is inherently and essentially anti-Christian. This is not true. The Christian gospel is not the "culture religion" of pretechnological society. Nor is it a particular expression of some general religious spirit that will necessarily vanish when scientific technology has done its work. The Christian gospel,

> rather, is the word of God calling men in every age and within any social or cultural ethos to take responsibility for himself and his neighbour in history before the Living God.[15]

That the Christian should embrace the modern world and affirm its values is almost universally accepted by theologians today.[16] Nothing epitomizes the modern spirit more than the claim that we have "come of age" and should not resist the secular outlook of the modern world. The phrase "man come of age" was, of course, used by Dietrich Bonhoeffer, the spiritual father of secular theology.[17] From his Nazi prison cell in 1944 Bonhoeffer wrote a series of letters to a friend, in which he spoke enigmatically of no longer needing God as a "working hypothesis":

> The movement beginning about the thirteenth century . . . toward the autonomy of man . . . has in our time reached a certain completion. Man has learned to cope with all questions of importance without recourse to God as a working hypothesis. In questions concerning science, art, and even ethics, this has become an understood thing which one scarcely dares to tilt at any more. But for the last hundred years or so it has been increasingly true of religious questions also: it is becoming evident that everything gets along without "God," and just as well as before. As in the scientific field, so in human affairs generally, what we call "God" is being more and more edged out of life, losing more and more ground. . . .[18]

> God is being increasingly edged out of the world, now it has come of age.[19]

The publication of Bonhoeffer's letters in 1951[20] fanned the flames of an already incendiary theological situation. The changes which gave rise to modernity were seen by many theologians as having profound consequences for the Christian faith. But two world wars and the inter-war situation in Germany had tended to distract the attention of theologians away from the particular issue of modernity. After the Second World War, however, there was a renewed effort to think through what we have become and what this implied for Christianity. In particular, attention was focused on the question of our autonomy. Nietzsche had declared that we must now create our own horizons of meaning and value,[21] but as we have seen, the realization of this fact casts a dark shadow over humankind. There are those, however, who are not at all overcome by this darkness. This is perfectly epitomized by the opening sentences of Edmund Leach's famous Reith lectures:

> Men have become like gods. Isn't it about time we understood our divinity? Science offers us total mastery over our environment and over our destiny.[22]

In the theological world there are, then, those who, influenced by such as Dietrich Bonhoeffer, affirm that our autonomy is the essence of our humanity. The freedom to create values and meaning is not something Christians should be afraid of, but they should accept it joyously. The cross represents the death of God and our concomitant liberation. We must assume full autonomy in the world and eschew the "world-to-come." It follows from the emphasis on this world that norms of behaviour are to be found not in some heteronomous authority (such as revelation) but in the concrete experience of everyday life. Thus in being freed for an open future we are severed from history and tradition.

This affirmation of our autonomy and involvement in the world is nurtured by an all-pervasive optimism. God's gift of freedom is perceived as the freedom to embrace an open future, full of possibilities and potentialities. There are those, however, who caution that an open future means a future open to both good and evil. Without a sense of history and tradition, and especially without a sense of *tragedy*, we are in grave danger of creating a world which, instead of being joyous and life-affirming, is dehumanized and depersonalized.[23] They argue, moreover, that technology is not the logical outcome of the biblical disenchantment with nature. It is to the thought of two such thinkers that we now turn.

2. A Dissenting View: Technique and the Eclipse of Human Autonomy

Of all the critics of modernity, Jacques Ellul is perhaps the best known and least understood. He is not a "professional theologian,"[24] yet he has had a major impact upon the theological world. The author of well over a score of books, he writes with a power and industry which is quite formidable, and he continues to write even though he doubts that people will correctly discern what he is saying.[25]

As he is such a prodigious writer, to take one major theme and focus on it hardly does justice to the totality of his concerns.[26] Nevertheless, the theme we shall focus on -- technique and modern society -- is one which permeates almost all his works and for our purposes is most appropriate. It was, after all, his book *The Technological Society*[27] which first brought him to the attention of the English-speaking world. The bleak picture of modern society which is found in the book is so compelling that anyone who wishes to understand the society he or she is living in simply cannot afford to ignore it.

Ellul is convinced that to understand the present-day world, we have to understand the nature and role of technique.[28] Now most people, when they look around at the modern world, do not see all-pervasive technique, but only various techniques. The originality of Ellul lies in his argument that technique is an all-pervasive social fact:

> The term *technique*, as I use it, does not mean machines, technology, or this or that procedure for obtaining an end. In our technological society, *technique* is the *totality of methods rationally arrived at and having absolute efficiency* (for a given stage of development) in *every* field of human activity.[29]

This definition requires elucidation. By it Ellul is implying, first of all, that in the technological society reason governs *all* human activity. But while reason governs all human endeavour, it is also instrumental and hence has as its goal efficiency. Technique is a question of "human reasoning concerning action, of efforts directed towards simplification and systematization, and of a concern for efficiency."[30] The machine embodies these ideals of rationality and efficiency, for it is created to function efficiently and rationally without human error. But Ellul sees a dimension to technique which goes beyond the machine stage: technique governs every field of human activity. In other words, Ellul believes that the principles of machine technology have spread to every area of human activity:

> But let the machine have its head, and it topples everything that cannot support its enormous weight. Thus everything had to be reconsidered in terms of the machine. And that is precisely the role technique plays. . . . Technique integrates the machine into society. It constructs the kind of world the machine needs. . . . All-embracing technique is in fact the consciousness of the mechanized world.
>
> . . . Technique thus provides a model; it specifies attitudes that are valid once and for all. The anxiety aroused in man by the turbulence of the machine is soothed by the controlling hum of a unified society. . . .
>
> But when technique enters into every area of life, including the human, it ceases to be external to man and becomes his very substance. It is no longer face to face with man, but is integrated with him, and it progressively absorbs him. In this respect, technique is radically different from the machine. . . . The mechanization which results from technique is the application of this higher form to *all* domains hitherto foreign

> to the machine; we can even say that technique is characteristic of precisely that realm in which the machine itself can play no role. It is a radical error to think of technique and machine as interchangeable; from the very beginning we must be on guard against this misconception.[31]

It is at this point that Ellul's analysis becomes most controversial. For Ellul is maintaining that technique functions in our society as the source and standard governing all relationships. Modern society is, in fact, a vast monolith governed by technique. Any way of perceiving the world other than through the medium of technique is becoming increasingly difficult, if not impossible.

> Henceforth, there will be no conflict between contending forces among which technique is the only one. The victory of technique has already been secured. It is too late to set limits to it or put it in doubt. The fatal flaw in all systems designed to counterbalance the power of technique is that they come too late.[32]

Technique obeys its own predetermined laws of logic; it does not brook interference with these laws. Thus, it escapes our control because we define ourselves by technique and therefore see no choices other than those presented to us by technique.[33]

Although this aspect of Ellul's formulation is controversial, he does give us pause for thought. How often is it said that the crises of the modern world -- be they ecological or energy-related -- are merely symptomatic of an incomplete stage in technology and that the correction of the imbalance merely requires a more complete technology? There certainly does seem to be within technology itself a drive towards totality. There is indeed evidence for the view that we have ceded to technique undivided allegiance to its power in return for its security. By its very nature technique precludes us from changing its course in midstream:

> We are faced with a choice of "all or nothing." If we make use of technique, we must accept the specificity and autonomy of its ends, and the totality of its rules. Our desires and aspirations can change nothing.[34]

A criticism that can be made of Ellul at this point is that he is guilty of a confused description of technique. He claims that technique determines that every relationship revolves around the search for means without any end but means. But is not the efficient and rational society of which he speaks so constituted because it serves some end, even if only happiness and the alleviation of man's estate? Ellul is firm in his answer. Although originally mastery over nature through technique may have been undertaken for the relief of man's estate, technique has now become an end in and of itself.[35] Moreover, says Ellul, because of this essential characteristic, technique will be unable to sustain itself, and modern society, predicated as it is on technique, will eventually collapse.[36]

Ellul insists, however, that technique in itself is not evil; it is a manifestation of something much deeper -- the human "will-to-power." He believes that the Bible, properly understood, queries the very assumptions of this will-to-power and search for security. Technique is the contemporary manifestation of the principalities and powers,[37] to which modern man gives his allegiance because of blind desire and false understanding:

> I am not saying that technique is one of the fruits of sin. I am not saying that technique is contrary to the will of God. I am not saying that technique is evil in itself.[38]

> I have never attacked technology. On the one hand, I have attempted to describe the whole sociological problem of technology, with emphasis on my conviction that the benefits accruing from technology are well worthwhile. On the other hand, I have attacked the *ideology* of technology and *idolatrous* beliefs about technology.[39]

Ellul's analysis of modern society is, in fact, more subtle than many people realize. He does see the technological society as having novel features, but he also stresses that seen from a biblical perspective technique is merely a manifestation of sin. "Babylon, Venice, Paris, New York -- they are all the same city, only one Babylon always reappearing. . . ."[40] "As far as power is concerned, exactly nothing has happened since Genesis."[41] The meaning of the Fall in Genesis is basically simple: the relation between God and humanity was ruptured, and we committed -- and continue to commit -- sin by acting independently of God. Since the Fall we have not known the will of God as the true ordering principle of the world, and we therefore encounter the world as hostile and chaotic. We therefore seek to create our own order, and to do so we fall back on what Ellul calls "the order of necessity." The order of necessity is what succeeded the spontaneous and immediate relationships of God's creation prior to the Fall.

> The nature which had produced everything in abundance for Adam's sustenance and joy becomes an ungrateful and rebellious nature which resists man. . . . As a result, Adam must coerce nature, must conquer this nature which gives him thorns and thistles -- Adam will have to give up his wheat, his fruits etc. . . . Thus, Adam finds himself in a relationship of conflict; he gets the upper hand through the means at his

> disposal; that is to say, through technique -- which cannot be an instrument of love but only of domination.[42]

In this sense, then, humanity today is doing nothing new: our use of techniques to guarantee survival has been characteristic of our existence since the expulsion from the Garden. But our reliance on our own power, our own ability to control nature, can paradoxically only lead to our enslavement; herein lies the truly deep significance of the Fall. We cannot obtain freedom through the use of technique.

> Technique is the present source of man's enslavement. It is not simply that, of course. Hypothetically, technique could be wholly a cause of man's liberation, just as, hypothetically, the state could be the source of security and justice, and the capitalist economy could be the source of happiness and of the satisfaction of needs. But all that is hypothetical. In reality, the state, the economic structure, and technique have been sources of alienation.[43]

If, however, there is a sense in which humanity today is doing nothing new in the use of technique to guarantee survival, there is a sense in which the modern situation is unique. As we have already observed, Ellul believes that "the qualitative change in the proliferation of techniques went hand in hand with a qualitative change in relationships in terms of technique. The new relationships have no historical precedent."[44] Modern society has gone beyond the use of technology as a tool and even beyond seeing technique as a structure of society; it now sees rationality as the only definition of reality and seeks to transform everything else into its own image.[45]

As a Christian, Ellul sees his task as the elucidation of the nature of the technological society; he does not seek to offer the Christian specific guidelines of response.

> But I refuse . . . to offer up some Christian or prefabricated socio-political solutions. I want only to provide Christians with the means of thinking out for themselves the meaning of their involvement in the modern world.[46]

Ellul is seeking to question modern assumptions about such things as freedom, the maintenance of power, and the belief in a stable order. Like Augustine before him, Ellul claims that freedom is not rooted in the self but is only possible in the context of obedience. By obeying God we are truly free; all other quests for freedom can only lead to enslavement.

> This non-obedience expresses itself, then, in a closure, a closing! In its estrangement from God, the world withdraws within itself and recognizes only its own imperatives: it is closed to all outside influences. . . . It entraps man in

situations without exit, all the while promoting itself as an all-fulfilling, all-encompassing system.[47]

Humanity today, in seeing the domination of nature as the guarantee of freedom, is in grave error.[48]

Ellul's vision of the technological society is a sombre one. He sees a world increasingly dominated by technique, a world in which real choices and options are constantly being diminished and in which we have become enslaved by our own devising. This blighted vision of modernity is also shared by George Parkin Grant, who agrees that "technique is ourselves."[49] Whereas much of Ellul's critique of modernity is biblically based, Grant's is philosophical.

Grant believes that the assumptions of modernity will ultimately lead to the destruction of many human rights which have been our inheritance. He argues in his little book *English-Speaking Justice*[50] that, as behaviour modification, genetic engineering, and birth control by abortion become more and more evidently manifestations of the mastery of human nature which is of the essence of the technological society, the conflict between the vision of the good, which is grounded in classical antiquity and Christianity, and the claims of convenience issuing from technology becomes sharper. Grant believes that eventually the technological enterprise, because it is grounded in different assumptions, will destroy the classical view of what is human, and a "terrifying darkness" will descend upon modern justice.[51]

The crisis of modernity lies in the fact that the technological society is said to be rational, and yet reason is said to be unable to validate the highest claims of life; it cannot tell one how to live. The moral pluralism of modern society implies, in fact, that we pursue means through rationalized technique and at the same time make questions of ends lie beyond rational discourse. How this came to be is the substance of Grant's first major publication, *Philosophy in the Mass Age.*[52]

In this book Grant points to a unique fact about North America -- it has no history before the Age of Progress. This means that North America is the very incarnation of modernity. It also means that it is very open to both good and evil, because of its lack of tradition. Its assumptions are those of modernity, because it knows no others. Only when we look at ancient classical cultures do we see (by comparison) that modernity is grounded in a very particular view of the world. Grant does not examine specific thinkers but takes a more holistic approach. He examines the view of classical antiquity on such matters as natural law, history, and freedom and compares it with the modern view on these issues.

Natural law maintains that:

> There is an order in the universe which human reason can discover and according to which the human will must act so that it can attune itself to the universal harmony. Human beings in choosing their purposes must recognize that if these purposes are to be right, they must be those which are proper to the place mankind holds within the framework of universal law. We do not make this law, but are made to live within it.[53]

It is the assumptions of natural law which bring it into conflict with modernity. Natural law assumes that the universe is a cosmos, not a chaos, and that this harmony and purpose may be discerned by the human being through reason. The goal of education is wisdom--the discovery of purpose in the universe. This is quite different from the modern view of reason as instrumental.[54] To live according to natural law was the supreme good of the human being. It was natural law in particular which came under attack when such as Nietzsche asserted that our essence lay in our historicity. As historical beings, it is claimed, we make our own laws and values; we are no longer subordinate to natural law. Similarly, the idea of freedom has undergone a change. Freedom is seen no longer as the ability to discover the true purpose of the cosmos and live in accordance with it, but rather as the ability to get what one wants when one wants it. Freedom in the modern world is indissolubly linked with our historicity -- we are historical creatures whose freedom lies in making the world as we will. It is at this point that we come face to face with the terrible question: Are there no limits to our freedom? Is humanity today unable to regard "wrong" as a meaningful category? Are we able to think only in terms of a "cost/benefit" morality? It is the inability to regard "wrong" as a meaningful category which will lead to our being engulfed by a "terrifying darkness," says Grant.[55]

Grant believes that an analysis of modern education shows his fears to be well founded.[56] One characteristic of modernity is, as we have seen, the belief that the mastery of chance is the chief means of improving the race. Grant notes why the modern university is particularly suited to cultivating those disciplines which issue in mastery of human and non-human nature.[57] The quantification/behavioural-oriented sciences are wonderfully appropriate for administering the controlling techniques necessary for modernity. In order for these sciences to function unimpeded, however, it is necessary to subscribe to the fact/value distinction. According to this view, previous accounts of reality confused normative and factual statements. The social scientist, however, gives an "objective" account of reality because he distinguishes facts from values. What this view does, of course, is dichotomize the scientific world, with its "objective facts," and the world of values, which are subjective because they are created by human beings. Once "good" and "bad" are perceived as subjective preferences, one possible brake is removed from the "triumphant chariot of technology."[58]

> The "value-free" social sciences not only provide the means of control, but also provide a large percentage of the preachers who proclaim the dogmas which legitimize modern liberalism within the university. At first sight, it might be thought that practitioners of "value-free" science would not make good preachers. In looking more closely, however, it will be seen that the fact/value distinction is not self-evident, as is often claimed. It assumes a particular account of moral judgement, and a particular account of objectivity. To use the language of value about moral judgement is to assume that what man is doing when he is moral is choosing in his freedom to make the world according to his own values which are not derived from knowledge of the cosmos. To confine the language of objectivity to what is open to quantifiable experiment is to limit purpose to our own subjectivity. As these metaphysical roots of the fact/value distinction are often not evident to those who affirm the method, they are generally inculcated in what can best be described as a religious way; that is, as a doctrine beyond question.[59]

The fact/value distinction is not the only dogma which we find dominant in the modern university; it is also dominated by historicism, together with its concomitant philosophical trappings. Historicism is, to use Grant's definition, "a belief that the values of any culture were relative to the absolute presuppositions of that culture which were themselves historically determined, and that therefore men could not in their reasoning transcend their own epoch."[60] Historicism is the inevitable consequence of accepting unequivocally the views of the men we have discussed in chapter three, according to whom history had uncovered the genesis of our beliefs, actions, and thoughts.[61]

In the present age, then, most of our assumptions are governed by technology, and most of our needs are created by it and for it. Modern education -- particularly in the university -- promotes general support for the assumptions of technology: "The university curriculum, by the very studies it incorporates, guarantees that there be no serious criticism of itself or of the society it is supposed to serve."[62] There thus reigns an educational tyranny which severs us from our past and augurs a bleak, dehumanized future.

Grant and Ellul provide an eloquent dissenting view to the idea that the technological human is autonomous. They believe that technique is eclipsing human autonomy; they do not believe that as modernity unfolds we will attain fuller humanity. It behoves us, therefore, to re-examine this issue in the light of what we have said in chapters one and two. Before we do that, however, we must summarize our argument about the nature of modernity.

Conclusion to Part II

Understanding the Phenomenon of Modernity

The argument of the preceding pages has been that it *is* meaningful to speak of modernity, that it does have a distinctive identity, and that its origins lie in the Scientific Revolution and the thought of such as Darwin, Freud, Marx, Machiavelli, and Nietzsche. There is, in fact, general agreement about the elements of modernity, but there is disagreement about whether it is salutary for humanity or not.

Modernity is constituted of four key elements: contingency, relativism, temporality, and autonomy.[1] We have seen how the Scientific Revolution eventually led to an understanding of the world in which causes are neither necessary, rational, nor purposive -- that is, an understanding of the world as contingent. We further observed how the social scientist and psychologist seek to explain human beings in terms of the nexus of their social relations and their historical self-understanding. This way of looking at the world is dominated by relativism and historicism. Nietzsche, it will be recalled, was the first thinker to really tackle the question of what our historicity and the finality of our becoming (our temporality) means. Finally, we saw how modernity lays great stress on the autonomy of humanity and our ability to make the world as we will.

It is the question of our autonomy which is the focal point of the debate about technology.[2] Technology is the fusion of knowing and doing. Whereas in ancient times science was knowledge of nature, in the technological age this knowledge is translated into power over nature. Such as Ellul and Grant argue that this enterprise can only lead to our dehumanization. We have exchanged the tyranny of nature for the caprice of our own wills, for in a pluralist society there is no agreement about ends. The "open" future is one which is open only to technological novelty; the choices presented to us are of this or that particular technological innovation. Technology itself is placed beyond question. To embrace technology is to accept the autonomy of *its* ends; we thus lose our freedom in the technological society.

There are those, on the other hand, who claim that technology has brought with it autonomy, and this autonomy is the means whereby we may realize fuller humanity. Biblical theology, properly understood, "disenchanted" nature and so prepared the way for our dominance of nature through technology. Technology is not only an expression of our autonomy, it is an expression of the knowledge and power of "man come of age." The Christian must distinguish between secularization and secularism, embrace the former and shun the latter.

As there is such disagreement about what modernity means for humanity, the issue requires further analysis. This is attempted in the following section, where particular emphasis is given to the relation of Christianity to modernity.

PART III

CHRIST AND MODERNITY

Chapter Five

Some Basic Issues

1. Continuity and Discontinuity in Christianity and Modernity

In Part I we discussed how the phenomenon of Christianity is centred upon Jesus Christ. Its essence cannot be grasped by rendering it into a rationally articulated statement. Prior to all articulation of the Christian faith is the experience of Christ, which is not solely individual and private (a simple pietism) but manifests itself concretely in the sharing of the experience with others in the Christian community. From the beginning, to be a Christian was to be part of a fellowship.[1] We argued in chapter two that only by grasping the nature of the early Christian community could the true nature of theologizing be understood. Theology arises out of the whole life of faith; its clarification and interpretation can only be understood within the context of worship and practice; that is, within the context of the *lex orandi*.

In chapter three we discussed how the beginnings of modernity were rooted in the break with the traditional (classical) views of the immutability of human nature and a fixed natural law. Human beings today see themselves as autonomous and living in a dynamic, complex society which is open to the future. There are no simple, heteronomous[2] solutions to the complicated ethical issues of today: abortion, population control, economic growth and development, the power of multinational corporations, environmental pollution, etc.

The significant point of these observations lies in the fact that in our description of early Christianity we found characteristics not dissimilar to those found in modernity. In the quest for normative self-definition, the early Christian did not appeal to some abstract principle or norm. Normative abstractions tend to result in uniformity, rigidity, and sterility. The early Christians, by focusing upon a concrete model -- Christ -- were able to adapt their behaviour and their beliefs according to the situation and circumstances of a particular time. Jesus Christ is the basic model for the Christian, not in the sense of imitation (as we pointed out in chapter one) but in the sense of embodying in himself an attitude to life and a practice for life.[3]

Thus, Jesus functions as an adaptable model -- one that can be realized in a variety of ways and in a variety of historical circumstances. Following Jesus is thus a dynamic activity, one in which the Christian is not *informed* in Christian living but *formed* in Christian living through the experience of Christ in the community.

The Christians of the first few centuries were obviously in a situation quite different from that of Christians today. The undifferentiated consciousness of Christians of the early centuries makes a comparison between them and us very difficult. But Christians of today, by virtue of the fact that they are Christians, stand in some kind of continuity with their fellow Christians of the first few centuries. The problem is how to delineate the lines of that continuity.[4] This, as we have seen, was essentially the question which Bauer and Turner were addressing. Admittedly, they were not speaking of the continuity between Christians of the first centuries and ourselves; rather, they were examining whether there was continuity between Christians of the fourth and Christians of the second century. Yet the problematic is the same, and Turner's approach, as described in our second chapter, is an instructive attempt to characterize Christianity. We have already stressed the importance of Turner's contention that the attempt to articulate normative self-definition arose out of the *lex orandi*. In two other respects the approach of Turner is very significant.

In the first place, Turner's view that the development of Christianity is best described as the interaction between fixed and flexible elements is one which has far-reaching consequences. Given that Christianity has expressed itself quite differently in different cultures and in different times, Turner maintains that there have always been elements of fixity within the Christian tradition. Turner is, in fact, supposing that it is meaningful to speak of Christianity as having a distinctive identity. If we are to speak of "Christianity" or "the distinctively Christian," this kind of presupposition simply must be made explicit; otherwise, only confusion will ensue, as there are those (some of whom we will discuss shortly) who assume that Christianity has no substantial identity, that it is a syncretism or, more exactly, an ongoing multiplicity of interpretations with family resemblances.

This leads to the second point. Turner's description of early Christianity is more convincing because he has taken account of *development*. Essentially, Turner points to the fact that it was in the tradition of the Church that the process of transposition was vitally realized before the historical consciousness brought this to light in a reflectively intentional fashion. If we do not take account of development, then, like Bauer, we will see Christian history as essentially discontinuous. By arguing for a *dynamic unity* of Christian development, Turner argues for what Lonergan calls "the unity of a subject in process." It is the interaction of the fixed and flexible elements which gives rise to development. To abandon the fixed elements is to neglect origins and make way for "enthusiasm" and the fanciful. To abandon the flexible elements is not to take development through history seriously, and an anachronistic "classicism" holds the field.[5]

When dealing with the debate on the nature of modernity (chapter four), we noted that George Grant was one of the most trenchant critics of modernity. When we look at his method of approach in discussing the issue, we find that it has a structural similarity to that of Bauer's discussion of early Christianity. Grant settles on the elements of the classical Greek view of natural law. He compares these to the view of nature and freedom found in modernity. He finds them to be incompatible and concludes that there has been a calamitous fracture in the Western tradition. Grant has taken a line of approach remarkably similar to that of Bauer. He is, in fact, fostering what we have called an "anachronistic classicism." Someone like Macquarrie, on the other hand, defines natural law as a "constant tendency" or an "inbuilt directedness." By defining natural law in this way, Macquarrie, like Turner, does take account of development. He is thus able, unlike Grant, to accommodate natural law to the modern world. He says, for example:

> Natural law is, as it were, the pointer within us that orients us to the goal of human existence. Actual rules, laws, and prohibitions are judged by this "unwritten law" in accordance with whether they promote or impede the movement toward fuller existence. Natural law changes, in the sense that the precepts we may derive from it change as human nature itself changes, and also in the sense that man's self-understanding changes as he sharpens his image of mature manhood. But through the changes there remains the constancy of direction.[6]

We have seen in chapter two what Turner considers to be fixed elements and what he considers to be flexible. Among the fixed elements are the "religious facts" (grasped through the *lex orandi*), the biblical revelation and the rule of faith. Among the flexible elements are "differences in Christian idiom" (resulting from differing contexts of time and place) and the individual personalities of the theologians. It is easy to see how the flexible elements influence the articulation of the Christian faith. Influenced by existentialism, Bultmann's articulation of the Christian faith is different from that of, for example, Schweitzer. Yet others view Christianity as so flexible that they dispense with traditional categories altogether -- including God.

Traditional forms, they say, no longer have any authority for the Christian.[7] Their view of Christianity seems to be that view we described as an ongoing multiplicity of interpretations with family resemblances. Christianity has no intrinsic identity of its own. The problem with this view of Christianity is that it is difficult to discuss meaningfully a phenomenon without some agreed criteria of rational discourse. Christianity becomes so flexible that it ceases to have any shape of its own -- for one person it is this, for another it is that. The understanding of Christianity becomes so variable that the word ceases to have any meaning, for there is no public agreement on its characteristics. Too high a degree of flexibility in the articulation of the Christian faith is to be avoided as much as the classical view of Eusebius, which saw Christianity as fixed and stable.

It is fairly easy to identify the flexible elements in Christianity, but perhaps not quite so easy to identify those that are fixed. A primary element of fixity lies in the belief in God the creator:

> Belief in God as a Sovereign Father of creation which is His handiwork forms an essential part of the basic realities of the Christian Church. His being may at times be described in terms more appropriate to the static and transcendent Absolute of Greek metaphysics, His fatherhood too closely approximated to mere causation, His Providence defined in terms drawn from the Hellenic concept of Pronoia. The religious fact still underlies the changing categories under which it is expressed.[8]

This particular element of fixity has important implications, as we shall see shortly.

Another element of fixity lies in the givenness of God. Although God gives of himself decisively in the Christ-event, this does not exhaust his givenness. The doctrine of the Holy Spirit was an effort to articulate the conviction that God not only has given but continues to give of himself in Christ. This givenness of God is what Turner calls a "religious fact," and it has certain implications. It precludes a docetic Christ, an infallible Church, and an ahistorical Bible.[9] It precludes the ahistorical view of Christianity (as in Gnosticism), and it precludes the classicist view (as in the Judaizing heresy). Christianity is not suprahistorical, nor is it time-bound and immutable. To say that Christianity is a dynamic interaction between fixed and flexible elements is thus a view of Christian development which excludes positions of extreme openness and also positions of extreme rigidity.

Grant and Ellul are two Christians who both reject the idea that modernity is salutary. They do so from two different perspectives.

Grant's rejection is primarily philosophically based, while Ellul's is biblically based. Despite these different perspectives, their assumptions are basically similar. Both men believe that the continuity of the Western tradition has been authentically broken. They seek to recover an horizon of meaning before the break: in Grant's case it is the classical Greek tradition and in Ellul's it is the biblical one. It is possible to argue that there has been a radical break in the Western tradition, but also to adopt the horizon of meaning of modernity and to reject previous horizons.[10] There are thus two polarized views on the relationship of Christianity to modernity. Gogarten's discrimination analysis is a mediating position between these two extremes. While acknowledging the dangers of secularism, Gogarten refuses to attack the present technological society in the name of Christianity. He maintains that the essence of Christianity lies in its historical understanding of the relationship of God to the world and to humanity. But if we are to exist historically, we must be responsible for our own destiny. Hence our autonomy and creative freedom towards the world is quite in accord with Christian faith. The subtlety of Gogarten's approach lies in his insistence that works are not to be invested with religious meaning; the gospel distinction between faith and works means that the Christian should attach no saving significance to works. We are thus free to preside over the world and govern it according to reason. Now it is true that Gogarten stresses that we receive the world "as an inheritance," and he does speak of our being a unity of "receptivity and activity." We shall suggest in our conclusion that the receiving of the givenness of God is fundamental to the Christian faith. Nevertheless, the problem with Gogarten's thought is that he seems to sever faith from works, so that it becomes difficult to see how the faith of the Christian is related to his activity in the world. We must therefore discuss further what restrictions Christian faith places on our autonomy in the world.

2. The Autonomy of Humanity

Those who see freedom simply as the pursuit of natural impulses and the shaking off of all restraints are not, of course, free. They are subject to the dictates of their passions. All things are lawful, says Paul, but not all things are helpful, and the Christian is not to be enslaved by anything (I Cor 6:12; cf. I Cor 10:23). Those who would be truly free are those who do not live solely for themselves: "For whoever would save his life will lose it; and whoever loses his life for my sake and the gospel's will save it" (Mk 8:35). Only those who do not live for themselves will become truly aware of themselves. To be free in Christ means to be free *from* alienation and self-centredness and free *for* love and service. Authentic Christianity seeks to impose God's will not heteronomously but theonomously. One finds one's true freedom and fulfilment in living for Christ. Christianity does not take away one's autonomy; freedom is central to the Christian message, as the Grand Inquisitor saw so clearly. Unlike the Grand Inquisitor, the Christian should not fear the freedom offered by Christ. It is not a freedom to do whatever one wants (an unprincipled libertinism), nor is it a freedom to create the world as one wills (the liberal ideal). It is a freedom which finds its fulfilment in Christ.

It is perhaps here that there emerges the widest gulf between the horizon of modernity and that of Christianity. We have seen how Machiavelli attacked the "ought" of moral discourse and said we should begin where we are, not with where we ought to be. Nietzsche furthered Machiavelli's thought by asseverating that God is dead: there is no absolute good. The liberal ideal of freedom is the logical outcome of such views: there are no constraints on our freedom; we will our own values. Thus, for Grant modernity poses a terrible threat to humanity, because we have lost the ability to think of "wrong" as a meaningful category, or to put it differently, we do not accept the unconditional ethical demand. Kueng describes well the elements of this unconditional demand:

> The unconditional ethical demand, the unconditional "ought" can be substantiated only in the light of an *unconditioned reality* (which, of course, cannot be proved by pure reason): an *absolute* which can impose a meaning and which cannot be man either as individual or as society. Any demand based -- for instance -- only on the human fellowship of communication and argumentation remains hypothetical in the sense that it presupposes a will to participate and so the human "ought" is ultimately deduced from a human "will." It is therefore a hypothetical imperative, based on interest.[11]

Throughout the history of Christianity there has been the temptation to turn this unconditional demand into an inflexible norm. It is only concretely that the unconditioned can be actualized -- that is, it is only in conditioned human existing that the unconditional demand can be embodied. We saw in our discussion of orthodoxy (chapter two) how a definition of orthodoxy has to be specific enough to meet real issues and flexible enough to apply to different times and places. Otherwise, definitions of orthodoxy become tautologous and confused. Similarly, when speaking of the unconditioned demand, one should beware of a tautology such as "You must never kill unjustly." This is saying that unjust killing is always unjust. It does not tell one whether it is unjust to kill, for example, a human foetus. The unconditioned demand has a validity only when applied in concrete situations. As each situation varies, so the application of the demand varies. This is where a genuine Christianity has such an advantage over legalistic religion and abstract moral philosophies. The Christian religion centres on Christ, who did not lay down universally binding rules but showed in his own life how the unconditional love of God might manifest itself in concrete situations. Jesus said that man was not made for the Sabbath but the Sabbath for man (Mk 2:27). By this he meant that normative rules of conduct are valid not for their own sake but for the sake of realizing the "greater good." The unconditional demand is always linked to the variable of the situation.[12] Grant's notion of natural law does not seem to be able to accommodate itself to this,[13] whereas Macquarrie's does.

The Christian is autonomous. But the Christian view of autonomy is different from that of modernity. The Christian view is inextricably bound up with belief in God. It is therefore undergirded by a *basic trust*.[14] For the Christian, theonomy is the "*condition of the possibility of the moral auto-nomy* of man in secular society."[15]

To be truly human is to be truly free (cf. Gal 5:1). In the Garden of Eden Adam and Eve were given the freedom of choice; they could either live in obedience or disobey and eat the fruit of the forbidden tree. In this way the Paradise story focuses on what it is to be human. Only the person who has freedom is truly human, for only decisions made in freedom can give focus and direction to one's life. Technology may be seen as our attempt to regain the hedonic existence which prevailed before the Fall.[16] The Christian should resist such attempts where they fail to provide a higher measure of freedom. In Huxley's Brave New World, technology has freed human beings from adversity, from "the sweat of his face" (Gen 3:19), but they are not truly free because they have been relieved of making decisions which lead to a genuine human authenticity. This is undoubtedly why such writers as Ellul see technology as such a major threat. They fear that the price technology exacts in its attempt to recover the idyllic existence of the Garden of Eden is the destruction of human freedom. Others, as we have noted, see our use of technology as the most authentic expression of our autonomy in the modern world. Moreover, such writers as Cox argue that there is an integral link between the technological enterprise and biblical faith. It is necessary now to look at this thesis in more detail, as it is a key issue in the wider discussion of the relation between Christianity and modernity.

3. Technology and Christianity: Some Biblical Reflections

The basic argument about the relationship of Christianity to technology is neatly summarized in the article by Cox entitled "The Christian in a World of Technology,"[17] where he maintains that "the cultural impact of biblical faith provides a necessary precondition for technology."[18] He then discusses the three main arguments which recur throughout the literature on the topic: that the biblical view of creation "disenchants" nature; that the Bible attributes a special worth to human work; and that the Bible not only allows for the possibility of changing things, but encourages such activity.[19] We will examine these three arguments in detail in view of the fact that they are frequently cited by those who see technology as the authentic expression of our autonomy in the world.

The first argument states that the biblical doctrine of creation "disenchants" nature and dichotomizes the natural world and the object of faith, thus creating a "cultural prerequisite" for technology.[20] In other words, the argument is that a magical world-view inhibits technology, which will make little headway as long as "forests and streams, fields and rocks are experienced as the locus of deity."[21] This argument is not as forceful as it first appears. It does not imply a *causal connection* between a disenchanted world-view and the rise of technology. It merely states that technology could not have arisen had not the disenchantment of nature first taken place. Moreover the kind of magical world-view envisaged as inhibiting technology would seem to be restricted to the most animistic of societies -- it presumably does not apply to, for example, Egypt where the pyramids (surely an advanced technological feat) were built.[22]

The second argument -- about the worth of human work in the Bible -- is equally superficial. In ancient Greece, it is claimed, only the slaves did manual work, and the free pursued the intellectual life. The material world for the ancient Greek was to be differentiated and divorced from the spiritual world. Humans were spiritual beings trapped in material bodies. As intellectual activity was concerned with higher, spiritual things, the physical world was disparaged. One of the reasons the ancient Greek demurred from the application of science was because it would involve physical labour, which was the domain of the slave. Hebrew society, on the other hand, had no such disparagement of physical labour; the human was not a spiritual being trapped in a physical body, but a psychosomatic unity.

> The biblical view of work is important because it provides one of the key elements in the intellectual structure of modern technology. When science consists of mere speculation and deduction, as it did with most of the Greeks and with the early Renaissance scientists, "modern science" has not yet appeared. Only with the utilization of equipment, with measurement, observation, experiment and laboratory work do modern science and technology appear on the scene. . . . An acceptance and respect for human work is an essential element of technology. Without the biblical vision, it might never have been possible.[23]

But this argument is surely oversimplified, for it fails to note how the circumstances of work in the Garden of Eden were different from those pertaining after the Fall. Adam, it is true, is put into the Garden "to till it and to keep it" (Gen 2:15). But what kind of work was this? Was it productive and necessary work? Was it work directed toward the ends of mastery and utility as in technology? *Shāmar* (to keep, take care of) and *'ābad* (to till, to work) are general, rather neutral words. They contrast starkly with the words of 3:18f., where, because of the curse God has put upon the ground (3:17), humanity is now in conflict with nature. The earth will yield thorns and thistles (3:18), and Adam is told, "in the sweat of your face you shall eat bread" (3:19). He is, moreover, expelled from the Garden "to till the ground from which he was taken" (3:23). In other words, after the Fall a quite different situation prevails, one in which efficient and productive work is necessary to overcome the curse upon the ground. Work may, in fact, be seen as a consequence of humanity's sin.[24] The argument that in the "biblical vision" there is "an acceptance and respect" for work glosses over this important detail.

> The third argument concerns giving humanity dominion over nature and links us with God's creative and innovative activity. It is, of course, a fundamental premise of the technological enterprise that change is not only necessary and desirable, but possible. A cyclical or fatalistic view of history would preclude change. But we are governed by the Judeo-Christian view in which history is seen as the theatre of human response. It is the scene of unique and unrepeatable events, demanding decisive and accountable responses.[25]

There may be a few dissenters, but most would accept the argument so far. The next stage of the argument, however, links humanity to God's creative and innovative action. Yahweh, it is claimed, did things which were "utterly new and unprecedented"[26] and we share in this activity. Two texts are often cited here. The first is Gen 1:26f. where humanity is given dominion over the earth and told to fill it and subdue it. The second is Gen 2:19 in which God commands 'the man' to 'name' the animals. It is argued that this involves humankind in "the naming and controlling which constitute the creative process."[27]

The central issue which is raised by the argument here revolves around the question of whether biblical theology *encourages* technology or simply *permits* it. The argument began by suggesting that the possibility of changing things is a prerequisite of the technological society, that the desire for change is "natural" for us and would be impeded only by a fatalistic or cyclical view of history. But the gist of what followed is that biblical theology *encourages* technology, because we are linked with God in the creative process. This is positing a causal relationship between biblical faith and technology. It is this position which James Barr attacks in his important article "Man and Nature -- The Ecological Controversy and the Old Testament."[28] Barr argues on exegetical grounds that the Old Testament does not support such a causal relationship. There are, furthermore, historical as well as exegetical objections to the idea. Bertil Albrektson, for example, has challenged the view that the "religions of the Ancient Near East traced the divine powers in the recurring interplay of the forces of nature, whereas the Old Testament finds God revealed in unrepeatable events in history."[29] And why, if the biblical worldview encourages technology, was its real fruition limited to the last few hundred years? Why the lapse of several centuries -- centuries in which Christianity was so culturally dominant? The reply usually runs along the lines that medieval theology mistakenly wed itself to to Greek philosophy and this inhibited the rise of technology. During the medieval period the authentic biblical understanding of humanity and nature was eclipsed. This is an oversimplified answer which fails to do justice to the complexity of Christianity. From the beginning Christianity expressed itself through a fusion of Hebraic and Greek thought-forms. The Judaism from which Christianity was derived was itself permeated by Hellenism. M. B. Foster's influential articles[30] are often quoted in support of the argument that the "un-Greek" elements in modern science are derived from a Christian doctrine of creation. But one of Foster's main points is that we cannot arbitrarily expel Greek ideas from Christianity. He also pointed out that modern science could not have arisen without a developed mathematics -- something we owe to the Greeks, not to the Bible. One must beware of setting up a dichotomy between Hebrew and Greek thought in which the rise of technology is said to have been encouraged exclusively by the former.

The argument that technology has been encouraged by biblical faith is, then, questionable. Moreover, it is an argument which can cut both ways as far as Christianity is concerned. Writers such as

Cox see technological development as salutary; if, therefore, this has a biblical basis it reflects favourably on Christianity. In a famous essay, however, Lynn White[31] agrees that technology had a biblical basis, but for him this was not to the credit of Christianity. White sees the modern technological exploitation of nature as something to be deplored. He argues, like Cox, that the Christian view of creation disenchants nature and makes its exploitation possible. But, unlike Cox, White therefore concludes that Christianity bore a huge burden of guilt for the ecological crisis. Both White and Cox agree on the basic premise that there is a biblical foundation for modern science and technology but disagree in their evaluation of whether this link is to be evaluated positively or negatively. Both these positions are to be differentiated from that of Ellul, who has a very negative assessment of technology but who also argues that the Bible does not encourage technology.

To question the argument that the biblical world-view encourages technology is not to deny any connection between the rise of technology and Christianity. We have argued that it was the philosophical break with tradition which gave rise to modernity and its concrete manifestation, technology. The philosophy of the Western world is, of course, intricately bound up with theology. We do not wish to oversimplify the argument about the origins of modernity. It is a complex phenomenon in which the philosophical break with tradition gave rise to modern liberalism and humanism and entailed the critique of Aristotelian science and medieval theology. In the past few pages we have focused on the precise question of whether there is an *integral* link between biblical thought and the rise of technology. Such a precise question militates against over-generalization, while at the same time it relates to other themes (such as our autonomy) which we have already discussed. So let us discuss further the biblical basis for the argument that there is a inner connection between biblical thought and the rise of technology.

The crucial texts in the discussion of the biblical data are to be found in the two creation accounts in the first two chapters of Genesis: the Priestly account (1:1-2:4a), designated by the letter P, and the Yahwist account (2:4b-25), designated by the letter J. The two accounts differ significantly and have different emphases. We shall consider them separately.

The order of creation in the P account is as follows: on the first day God creates light and darkness (1:3-5); on the second day heaven and the waters (1:6-8); on the third day earth and vegetation (1:9-13); on the fourth day the sun, the moon, and the stars (1:14-19); on the fifth day birds and sea creatures (1:20-23); and on the sixth day land creatures (1:24-25) and then humanity (1:26-30). God rests on the seventh day "from all his work which he had done" (2:2). Some commentators have made a great deal of the fact that humanity is last in the order of creation. But such an interpretation overlooks one vital point: the creation of humanity is *not* the end of the P account of creation! As Claus Westermann notes:

> . . . Genesis I does not really close with the creation of man; rather, in 2:1-3, definitely a P section, something is introduced which actually does not belong to the work of creation, but which is given even stronger emphasis by being presented as the conclusion: man is *not* the goal of God's creation. From the very beginning the seven-day framework has been progressing towards the seventh day. The goal is really the solemn rest of that day. In the blessing and hallowing of the seventh day, we may detect the still veiled goal, the day of worship on which the responding congregation audibly utters the praise of the Creator which at creation was still implicit in God's own contemplation of his work.[32]

This is surely most significant. The focal point of the P account is God, not humanity, and creation ends with a day which is hallowed and reserved for the praise of the creator. The word used for "to create" (*bārā'*) is one which has only God as its subject.[33] Creation is God's alone.

In Gen 1:26ff. God gives us "dominion" over the earth:

> Then God said, "Let us make man in our image, after our likeness; and let them have dominion over the fish of the sea, and over birds of the air, and over the cattle, and over all the earth, and over every creeping thing that creeps upon the earth." So God created man in his own image, in the image of God he created him; male and female he created them, and God said to them, "Be fruitful and multiply, and fill the earth and subdue it; and have dominion over the fish of the sea and over the birds of the air and over every living thing that moves upon the earth."

This text is by far the most important in the whole debate about the relation of technology to Christianity, for commentators almost universally take it to mean that God hands over his creation to the human being, who thus enters into a partnership with God and "finishes" or "completes" the work of creation.[34]

But is such an interpretation really warranted by the text? There are two key questions in the interpretation of this passage. First, what does it mean to say that "man" was created "in the image of God?" Second, what exactly is the force of the command "to have dominion?"

The question of what the "image" or "likeness" of God means is an extremely controversial one. The phrase occurs in two other places in Genesis (both P passages) -- Gen 5:1-3 and Gen 9:1-6.

Gen 5:1-3 reads:

> This is the book of the generations of Adam. When God created man, he made him in the likeness of God. Male and female he created them, and he blessed them and named them man when they were created. When Adam had lived a hundred and thirty years, he became the father of a son in his own likeness, after his image, and named him Seth.

Gen 9:1-6 reads:

> And God blessed Noah and his sons, and said to them, "Be fruitful and multiply and fill the earth. The fear of you and the dread of you shall be upon every beast of the earth, and upon every bird of the air, and upon everything that creeps upon the ground and all the fish of the sea; into your hands they are delivered. Every moving thing that lives shall be food for you; and as I gave you the green plants, I give you everything. Only you shall not eat flesh with its life, that is, its blood. For your lifeblood I will surely require a reckoning; of every beast I will require it of man; of every man's brother I will require the life of man. Whoever sheds the blood of man, by man shall his blood be shed; for God made man in his own image."

The exegetical problems surrounding these passages are notorious, and the history of their interpretation is full of all kinds of suggestions as to what the image of God might be. The phrase has been taken to refer to reason, to immortality, to freedom, to actual physical likeness to God, and to human dominance in the world. It is not our intention to cover all the various possibilities here.[35] We will concern ourselves with the possibility that the "image of God in man" refers to humanity's dominion over nature. This is a common view, explicitly expressed in the words of H. Berkhof:

> When Gen i.27 says God created man in his own image, the whole passage i.26-28 makes it clear that what is mainly thought of is man's dominion over nature. As God is the Lord over the whole of his creation, so he elects man as his representative to exercise this lordship in God's name over the lower creation.[36]

Such an interpretation surely reads too much into the text. The point of Gen 1:26f. is a simple one: humanity was created like God. Certainly one may ask in what way we are like God, but it is no use looking to the text for an answer, because it does not tell us.

Moreover, as James Barr says, "It was probably essential to the writer's position that it could not be stated."[37] Barr himself believes that the passage is "concerned with a long-standing, and peculiarly Israelite, debate about the question of the likeness between God and man."[38] That the passage links the image of God in humanity to dominion over nature cannot be disputed. But this is a long way from claiming that the image consisted in dominion. To quote Barr once again: "It is rather likely to be a consequential relation: *since* man is in the image of God, let him have dominion etc."[39] In the other two passages the image is not linked to dominion. Adam has a son in his own likeness (5:3), and murder is forbidden because humanity is made in the image of God (9:6).

The best summary of the significance of Gen 1:26ff. comes from Claus Westermann:

> We cannot attempt to discover what in man is like God's image. Actually, the primary concern of the statement is not about man, but about the creative act of God. This makes it clear that man is thus set apart from everything else that is created; but this is to be understood only in the total context of the creation story. One cannot make out of this an assertion concerning the nature of man. It makes a difference whether one says, God made man after his image, or, man is the very image of God. . . . The statement really means that God has made man to communicate with him, so that man might speak to God and he might hear God's word. This description of man means . . . that man can maintain his humanity only in the presence of God. Man separated from God has not only lost God, but also the purpose of his humanity.[40]

Much has been made of the fact that the word usually translated "to have dominion" (*rādā*) is "strong." Von Rad maintains that besides meaning "to rule," it can also mean "to tread" or "to trample" (as in a winepress), and similarly that "to subdue" (*kābaš*) can also mean "to stamp."[41] But one should beware of making too much of this. In Joel 4:13 it appears as though the word *rādā* is used in the sense of treading a winepress, but elsewhere the verb means simply "to govern" or "to rule." Yet even if *kābaš* does suggest "stamping" or "trampling down," the meaning of words should not be determined in isolation; rather, one should also examine the context in which they are used. In the following verses (1:29ff.) we are told that the earth is to provide food for the animals; there is no eating of meat in the Garden. The animals live in communion with each other, without fear, and of all the animals human beings are pre-eminent. If we compare these verses with the covenant with Noah (9:1-3, a P passage), there is a remarkable difference. Here, as in 1:28a, God commands, "Be fruitful and multiply and fill the earth," but then adds, "the fear of you and the dread of you shall be upon every beast of the earth," (9:2) and the animals are handed over for food. The original harmony of the Garden has been disrupted,[42] and humankind not only has "dominion" over the animals but induces fear in them and uses them for food.

Those who see in the commands "to subdue" and "to have dominion" the justification for technology do not, then, do full justice to the context of the commands. The P account stresses the sovereignty of God, not of humanity. To suggest that God hands over his creation for us to *exploit* or *enslave* simply reads too much into the text of 1:26ff. The domination of humanity which leads the world of nature to fear us only appears *later* in the covenant made with Noah.

When we turn to the J account of creation, however, it has to be admitted that we have a much more anthropocentric account than that of P. The order of creation in J stresses importance of humanity. As Von Rad comments:

> Whereas in ch. 1 creation moves from the chaos to the cosmos of the entire world, our account of creation [J] sketches the original state as a desert in contrast to the sown. (Thus: ch. 1, chaos-cosmos; ch. 2:4bff., desert-sown.) That is the great difference from ch. 1. It is man's world, the world of his life (the sown, the garden, the animals, the woman), which God in what follows establishes *around man*; and this forms the primary theme of the entire narrative, *ādām ʾadāmā* (man–earth).[43]

It is, moreover, in the J account that God brings the animals to 'the man' for him to name (2:19f.).

Once again, however, caution needs to be exercised lest we read too much into the text. To take 2:19f., first, one might agree that naming is the first phase of objective knowledge and that with the classification of animals and species we have the beginning of scientific inquiry. But this is a long way from deriving *technology* from the text. Technology is a fusion of knowing and doing; it is the application of scientific knowledge in an attempt to control nature. Scientific inquiry in itself does not necessarily lead to technology.[44] As for the anthropocentricity of the J account, it should be pointed out that in it there is no mandate similar to the one "to have dominion" found in P. Humankind is put into the Garden "to till it and to keep it" (2:15); there is no suggestion of domination. Indeed, the J account makes it quite clear that it is only after the Fall (Gen 3) that we have the beginnings of technology with the building of cities (Gen 4:17) and the making of tools (4:22). It is only after the Fall, when the earth no longer yields its fruit freely (3:18f.), that technology is needed with its values of mastery, efficiency, and utility.

We have treated the J and P accounts of creation separately because we believe that examined individually they do not support the argument that biblical theology encourages technology. It is only by combining the anthropocentricity of the J account with a dubious interpretation of the "dominion" texts of the P account that one can find any support for such an argument. It is true that it is only in recent times that the J and P accounts have been separated in biblical studies. But this in no way invalidates what we have said, for there is a clear distinction to be made between the historical argument about the relation of Christianity to technology and the exegetical one. One may argue that in the past the Judeo-Christian doctrine of creation, God, and humanity was understood in such a way that it encouraged technology. This is an historical argument which may only be settled by examining the historical evidence. The exegetical argument is that biblical theology, properly understood, encourages technology. This is an exegetical question which must be settled on exegetical grounds, and this is what we are attempting to do here. It is therefore quite legitimate to make use of the results of modern source-critical studies of Genesis. Such studies have revealed that each of the two creation accounts J and P has its own point of view. But neither viewpoint supports the argument that a doctrine of creation encourages technology.

If one wished to take a more "holistic" view of the Bible, and leave out of account the differences in the various sources, or even differences between the various books, one would still find a view of creation consistent with our argument. What references there are to nature[45] suggest not that we are set apart so that we may dominate it, but rather that we are a part of it and have responsibility for it. We are creatures of God and responsible to God; we must render an account before God (Rom 14:10). We are God's *vice-gerents*,[46] a useful word that suggests the idea of one given rule by a superior and thus ultimately responsible to that superior. Our dominion is a gift of God and must be exercised in the light of that knowledge. Psalm 8 is often quoted by those who wish to stress our absolute rule over nature. Now while in Ps 8:6-8 we are said to have been given dominion over nature, the preceding verse (8:5) says that we are "a little less than God,"[47] and the psalm ends, not with a eulogy of humanity, but with praise for God the creator of all (cf. Ps 104). The emphasis falls not on our autonomy, but on God's creation.

It is unfortunately the case that too often in Christian theology we have been separated from nature. In neo-orthodoxy "nature was usually treated as the setting for the drama of human redemption. In existentialism, nature was the impersonal stage for the drama of human existence."[48] Since the time of Augustine, theology has tended to show an indifference to nature by focusing either on "the world to come" or, more recently, on our historicity in the world. Lynn White, however, goes too far in his argument that Christianity is to blame for the ecological crisis. The historical evidence indicates that the technological exploitation of nature owes more to the liberal concept of freedom than it does to a Christian doctrine of creation. Certainly the biblical view of humanity and nature does not give grounds for uncontrolled exploitation. It rather stresses our unity with nature.[49] We are indissolubly linked with the environment. In Genesis "God 'forms' him [man] from the ground; the bond between man and earth given by creation is expressed with particular cogency by the use of the Hebrew words *ādām ʾᵃdāmā*."[50] Moreover, nature is not just for our use but has value in itself. Nature, as well as humanity, will receive eschatological redemption (Rom 8:19ff.). God "delights" in his creation (Job 38-39; Ps 104; cf. Gen 1:31), and nature praises its creator (Ps 19, 89; Rev 5:13). Jesus tells us that God cares even for the sparrows (Mt 20:29). In Gen 9:10 God establishes a covenant not only with humanity but with all creatures, and Hosea sees the new covenant as one made not only with humanity but with the whole of creation (Hos 2:18). A world immersed in "the knowledge of the Lord" is a world in which the whole of nature lives in harmony (Is 11:6-9). The apocalyptists of later Judaism look forward to a time when *the whole of the created order* will be transformed in accordance with God's will.[51]

There are very specific regulations laid down in the Old Testament about the use of nature which quite clearly exhibit a respect and love for God's creation. Animals are to rest on the Sabbath as are human beings (Ex 23:12). The taking away of a bird with its eggs and the muzzling of a threshing ox are prohibited (Deut 22:6; 25:4). There are restrictions on cutting down trees (Deut 20:19-20), and the land is to lie uncultivated every sabbatical year (Lev 25). Such regulations militate against the kind of exploitation of nature which is a concomitant of the modern technological enterprise.[52]

The regulations we have just quoted are from the Old Testament. There is nothing similar in the New. The New Testament has very little to say about our use or misuse of nature. There are several reasons for this. The Old Testament contains a wider diversity of books dealing with all aspects of life (cf. the Wisdom literature) because it covers the religious history and culture of a people over a period of nearly a millenium. The books of the New Testament at most span a period of a hundred years. They are, moreover, addressed mainly to urbanized communities and centre almost exclusively upon the theme of personal salvation. The early Christians expected the imminent end of the world and consequently gave little thought to the wider questions involved in our relationship to nature. Paul does, of course, have a good deal to say about the freedom of the Christian. But his remarks are set in the context of the Jewish-Christian debate about the Law and freedom in Christ. We should be very cautious about applying Paul's words about the freedom of the son and heir (Gal 4) in a direct manner to the use of technology today. The question of how Christians are to understand their faith in a technological age is one we shall focus upon shortly. Before we do so, however, it will be useful to enucleate the results of our observations on the view that the Bible encourages technology.

There is a need for a basic reconsideration of the relation between Christianity and the technological enterprise. In our view the Bible does not encourage technology. But this is not to say that it *discourages* it. Even the severest critics of technology (such as Ellul) do not say this. The Bible may *permit* technology, even if it does not encourage it. But in the technological exploitation of nature it should never be forgotten that the use of nature countenanced by the Bible is a responsible one, one made in a spirit of harmony and receptivity. The liberal concept of freedom which sustains modern technology has as its concomitant values efficiency, mastery, and ultility. This has led to economic inequality, pollution, and depersonalization. There is surely a need for a radical change of attitude, and Christian theology can help to bring this about by showing that human autonomy is not at all incompatible with a nurturing and conserving attitude towards God's creation.

Conclusion

Christian Self-Understanding in a Technological Age

Christianity is a vital faith which has developed through the centuries and accommodated itself to various historical circumstances. It is not a faith, however, which can authentically accommodate itself to all circumstances. While an attempt at a recovery of a romantic pre-technological faith does not accord with a dynamic understanding of Christianity, neither does the indiscriminate embracing of the assumptions of modernity accord with the integrity of the Christian faith. To limit Christianity to a particular world-view is, as Gogarten observed, to conceive of it ideologically. The focus of Christianity is Christ; the lordship of Christ transcends all ideologies and cultures. The interaction between what we have called "fixed and flexible elements" in Christianity precludes both positions of extreme openness and positions of extreme flexibility in the articulation of the Christian faith. In other words, the Christian in the technological society must be discriminating in the approach to modernity; simple acceptance or rejection is too facile. The Christian should not, for example, reject the emphasis of modernity on human autonomy, for Christianity too emphasizes freedom. The task of the Christian is rather to point out that his or her faith gives autonomy a richness and meaning which modernity cannot. The Christian has the potential of existing in radical freedom, both "*from* himself and his world, because nothing is significant save for God's everlasting love for it, and *for* himself and others, because within the boundless scope of God's love everything is of infinite worth."[1] Modern secular notions of freedom are limited to the horizon of the finite. Thus they degenerate easily into dehumanizing ideologies. The Christian notion of a self-transcending freedom liberates because it is not limited in this way. The ultimate question posed by human autonomy is whether the ends of humanity are to be willed from within the horizon of the finite or not. The liberal concept of freedom asserts that this is the case. But for the Christian, freedom is realized only by choosing to live beyond the limits of the finite. It is received through the givenness of God, a "givenness which we could not create or imagine but only receive."[2] This receiving of the givenness of God has always been basic to the Christian faith. It is described in the following terms by Eugene Long:

> In every historical encounter there is . . . a residue, a mystery which is not capable of being examined by traditional frames of thought. Starting with man, we discover that man is ultimately a question to himself which admits of no answer out of man's creative rationality. Man is not primarily a doer but a receiver of the reality which is possible in each encounter. This reality comes by way of the other whom he encounters, yet it is neither himself nor the other. As such, we participate in it and are permitted to point to it as that which summons and addresses us.[3]

Modernity affirms our historicity in the world. Once again, the task of the Christian is not to reject this affirmation, but to point out that it is also emphasized in Christianity. But for the Christian our historicity does not mean that we cannot know what we are "fitted for" (to use Grant's phrase). In the incarnation God entered into the theatre of human activity and thus showed that his purposes were to be worked out in history. In the first centuries of its existence, despite Gnostic and Platonic influences, Christianity never lost sight of this important truth. Christianity is an historical faith. Christians should not be discomforted by the changing images of the human which have surfaced in modernity, for Christianity is tied neither to a static view of humanity nor to an immutable morality. To say that we are historical beings who determine our selfhood as we go along is not to plunge us into an ethical relativism, for in Christianity there is, if we may adapt Macquarrie's words, a "constancy of direction."[4] The direction may be grasped through various categories of thought (e.g., apocalyptic, ontology, existentialism), but the goal remains constant.

It is true that Christianity has often placed the emphasis on "the world to come" rather than on our existence in the world. Yet it is also true that Christianity has never been oblivious to the fact that we have physical as well as spiritual needs. The biblical account of creation, as we have seen, conceives of the human being as a psychosomatic unity. This biblical conception of humanity has several important implications. Such "mundane" matters as decent housing, sufficient food, good health, and education are all of great importance if we are to take the needs of the body seriously. A technology which genuinely provided for "the relief of man's estate" would therefore be salutary. In the contemporary technological society, however, the dispossessed have benefitted least from technology. This accentuates the need to renew the prophetic protests against economic and social injustice. Our technological society is simply not in accord with Christian ideals of charity and justice. It has promised much and delivered little to the poor of this world. In the midst of plenty there is poverty and starvation. Moreover, technological society's emphasis on consumerism has engendered a spiritual destitution of hideous proportions. The burgeoning "electronic culture" has spotlighted this in a macabre way. Vast populations of people spend hours each day passively watching television where they are nourished by such cultural masterpieces as *Three's Company, The Price is Right,* and *Dallas.* The vast profit potential of the video tape market has intensified the commercial appeal to humanity's baser instincts, exemplified in pornography with its shameless exploitation of women and children. Increasingly, alienation, madness, and violence have become the hallmarks of the technological society.

It is in the proliferation of nuclear weapons that we see most clearly the hallmark of our age. It is, of course, a suicidal competition for supremacy in weapons of mass destruction which has brought us to the edge of extinction as a species. But instead of seeking a way out of this dilemma by abolishing or reducing nuclear stockpiles, the "solution" is sought by driving towards an even more complete and sophisticated war technology. This "star wars" mentality epitomizes exactly what we noted in chapter four: problems in the technological society are almost always attributed to an incomplete technology; the solution therefore is not to dismantle it but to strive to make it ever more dominant and complete. Technology itself is not called into question.

The modern technological enterprise emphasizes progress based on profit and innovation. Business and military and scientific interests have thus overwhelmed the political process. No longer is the best political order seen as that which is most conducive to the practice of virtue or how we should live. Political choice is reduced to choosing between "brand-name" politicians.

The response of Christians to the injustices and alienation of the technological society varies. Some Christians, like those of the *Catholic Worker,* reject existing political and social systems and advocate a non-violent revolution to establish a social order in accord with Christian truth. Others attempt to reform and reorient the technological society from within the existing structures. We have stressed that the Christian faith does not allow a simple, unequivocal appeal to an immutable and external norm. There are bound to be differences among Christians over how to react to the pressing issues of the day. Christianity is not grounded in an heteronomous theology; it is grounded in Christ and guided by the Spirit working through the community of faith. It is in the context of faith and worship that the Christian comes to know the lordship of Christ. But there can be no meaningful religious life without commitment and involvement which are anchored in a "thankful thinking." We may well find the historical Jesus an elusive figure, but the Lord Jesus is ever revealing himself through that fellowship which binds all Christians together in the adoration of God. As Schweitzer put it so eloquently:

> Christianity is a Christ-mysticism, that is to say, a "belonging together" with Christ as our Lord grasped in thought and realized in experience. By simply designating Jesus as "our Lord" Paul raises him above all the temporally conditioned conceptions in which the mystery of His personality might be grasped, and sets him forth as the spiritual Being who transcends all human definitions, to whom we have to surrender ourselves in order to experience in Him the true law of our existence and being.[5]

Those Christians who believe that the ancient wisdom of Christianity is not fundamentally at odds with the assumptions of the technological enterprise would do well to ponder the sombre fears of Grant and Ellul. There are grave dangers in accepting the doctrine that the mastery of chance is the chief means of improving the human race. It seems almost inevitable that this will result in stressing the qualities of success, efficiency, and usefulness and that this is more likely to lead to dehumanization rather than to fuller humanity. Yet there is a great deal to the argument that, if we are being engulfed by the twilight, it is of our own making, and it is within our power to disperse it. A deterministic pessimism is simply not congruent with God's gift of freedom. The way ahead may indeed be fraught with difficulties, but the Christian does have certain basic convictions to use as guidelines. None are more basic than those contained in the words of John Taylor: "Technology is safe only in the context of worship, and science should walk hand in hand with sacrifice."[6]

NOTES

Notes to Chapter One

(1) ET: *On the Intention of Jesus and His Disciples,* in *Reimarus: Fragments,* ed. Charles H. Talbert, trans. Ralph S. Fraser (Philadelphia: Fortress Press, 1970).

(2) ET: *The Life of Jesus Critically Examined,* trans. from the 4th ed. (1840) by Marian Evans [George Eliot] (New York: Calvin Blanchard, 1860). Republished in 1970 by Scholarly Press, Michigan.

(3) ET: *The Quest of the Historical Jesus. A Critical Study of its Progress from Reimarus to Wrede,* trans. W. Montgomery (London: A. & C. Black Ltd., 1st ed. 1910, 2nd ed. 1936).

(4) Semler in *Beantwortung der Fragmente eines Ungenanten insbesondere vom Zweck Jesu und seiner Juenger,* quoted by Charles H. Talbert in his Introduction to *Fragments,* p. 1.

(5) The idea that Jesus was a political revolutionary has been most successfully popularized by Joel Carmichael, *The Death of Jesus* (New York: Macmillan, 1962). The most serious and scholarly treatments of the matter are found in Robert Eisler, *The Messiah Jesus and John the Baptist and other Jewish and Christian Sources* (New York: Dial Press, 1931); S. F. G. Brandon, *Jesus and the Zealots: A Study of the Political Factor in Primitive Christianity* (Manchester: The University Press, 1967). The best critique of the literature is found in Martin Hengel, *Was Jesus a Revolutionist?* (Philadelphia: Fortress, 1971).

(6) *Quest,* p. 23.

(7) W. G. Kuemmel, *The New Testament: The History of the Investigation of its Problems* (London: S.C.M. Press, 1973), pp. 90f.

(8) So stated by David L. Dungan in his review of *The Lives of Jesus Series* in *Religious Studies Review* 4 (1978), p. 264.

(9) See Kuemmel, p. 55.

(10) See the Introduction of Talbert to *Fragments,* p. 40.

(11) Quoted in Sir Malcolm Knox, *A Layman's Quest* (London: George Allen & Unwin, 1969), p. 34.

(12) *Fragments,* p. 64.

(13) *Ibid.,* p. 64f.

(14) Mt 28:16-20 was added later, claims Reimarus. The arguments adduced in favour of this assertion are now very familiar: (1) it appears nowhere else; (2) Jesus did not baptize (Jn 4:1f.); (3) Peter

would not have required the vision of Acts 10 if he had received from Jesus the commission given in Mt 28:16-20.

(15) Reimarus argues that Jesus never said, for example, "My kindgom is not of this world" (Jn 18:36).

(16) *Fragments*, p. 250.

(17) *Ibid.*, p. 245.

(18) *Quest*, p. 23.

(19) *Ibid.*

(20) *Ibid.*, p. 78.

(21) The following remark, for example, is directed against a work of Eschenmayer: "This offspring of the legitimate marriage between theological ignorance and religious intolerance, blessed by a sleep-walking philosophy, succeeds in making itself so completely ridiculous that it renders any serious reply unnecessary." Quoted in *Quest*, p. 98.

(22) *Life of Jesus*, Preface to first edition.

(23) The word "myth," of course, is susceptible of many interpretations, and the use of the word seems almost inevitably to lead to misunderstanding. See J. W. Rogerson, "Slippery Words V. Myth," *ExTim* 90 (1978), 10-14; also his *Myth in Old Testament Interpretation*, BZAW 134 (Berlin & New York: De Gruyter, 1974).

(24) *Life of Jesus*, Preface to first edition.

(25) *Ibid.*, pp. 48f.

(26) *Ibid.*, p. 42.

(27) *Ibid.*, pp. 64ff.

(28) See Kuemmel, *History*, n. 163.

(29) Schweitzer, *Quest*, p. 92.

(30) Cf. Stephen Neill, *The Interpretation of the New Testament 1861-1961* (London: Oxford University Press, 1964), p.16: "When Strauss had finished his critical work and given us Jesus as he understands him to have been, has sufficient account been given of what lies at the origins of a great world movement? . . . As Strauss understands it, Jesus lived on in the faith of his disciples, and this faith was strong enough to create the belief in his resurrection. But the kind of Jesus who is indicated in Strauss's pages was not the kind of person to create that kind of faith."

(31) John H. Hayes, *Son of God to Superstar: Twentieth Century Interpretations of Jesus* (Nashville: Abingdon, 1976), p. 30.

(32) Leipzig: Hinrich, 1900. ET: *What is Christianity?* (London: Williams and Norgate, 1901).

(33) Norman Perrin, *The Kingdom of God in the Teaching of Jesus* (London S.C.M. Press, 1963), p. 16.

(34) See n. 3 above.

(35) Schweitzer's own reconstruction of the historical Jesus was outlined in an earlier work than *Quest*: *Das Messianitaets- und Leidengeheimnis. Eine Skizze des Lebens Jesu* (1901). This was partly translated by Walter Lowrie in 1914 with the title *Mystery of the Kingdom of God* (London: A. & C. Black, 1914; Macmillan, 1950).

(36) For a balanced biography of Schweitzer, see James Brabazon, *Albert Schweitzer: A Biography* (New York: G. P. Putnam's Sons, 1975). See esp. his chapter "A Crack in the Myth," pp. 402ff.

(37) See A. Schweitzer, *Out of My Life and Thought: An Autobiography*, trans. C. T. Campion (New York: Holt, Rinehart & Winston, 1933).

(38) *Quest*, p. 238. It is unfortunate that Schweitzer used the word "eschatological"; "apocalyptic" would have been more appropriate. See Neill, *Interpretation*, p. 195.

(39) Johannes Weiss, *Jesus' Proclamation of the Kingdom of God*, trans. and ed. Richard H. Hiers and David L. Holland (Philadelphia: Fortress, 1971).

(40) William Wrede, *The Messianic Secret in the Gospels: A Contribution Toward the Understanding of the Gospel of Mark*, trans. J. C. G. Greig (Cambridge: J. Clark, 1971).

(41) *Quest*. p. 238.

(42) Quoted by N. Perrin, *Kingdom*, p. 18, n. 1.

(43) Some rejected an eschatological Jesus simply because he was so foreign; e.g., C. W. Emmet, *The Eschatological Question in the Gospels, and Other Studies in Recent New Testament Criticism* (Edinburgh: T. & T. Clark, 1911), says: "They [the liberals] portray for us a Christ whom we can unreservedly admire and love, even if it is a little doubtful whether logically we ought to worship him. The Jesus of eschatology it is difficult either to admire or to love; worship him we certainly cannot." Quoted in Perrin, *Kingdom*, p. 40.

(44) The question of whether Mark was the earliest Gospel has come under increasing attack in recent years. See esp. David Dungan, "Mark -- The Abridgement of Matthew and Luke," *Jesus and Man's Hope I* (Pittsburgh: Pittsburgh Theological Seminary, 1971), pp. 51-97.

(45) Even after Wrede, there are those scholars who still defend the idea that Mark contains a reliable historical outline of the life of Jesus. The best exponent of this view is C. H. Dodd. See in

particular his "The Framework of the Gospel Narrative," *ExTim* 43 (1932), 396-400, and, more recently, *The Founder of Christianity* (New York: Macmillan, 1970). On Dodd's article see D. E. Nineham, "The Order of Events in St. Mark's Gospel--An Examination of Dr. Dodd's Hypothesis," in *Studies in the Gospels*, ed. D. E. Nineham (Oxford: Blackwell, 1955), pp. 223-240. Reprinted in *Explorations in Theology I* (London: S.C.M. Press, 1977), pp. 7-23.

(46) See my own article "The Incomprehension of the Disciples in the Marcan Redaction," *JBL* 91 (1972), 491-500.

(47) *Quest*, p. 333.

(48) *Ibid.*, p.394.

(49) *Ibid.*, p. 353.

(50) *Ibid.*, p. 356.

(51) *Ibid.*, p. 363.

(52) Jesus thought of himself as the Son of Man who would be made manifest -- *ibid*, p. 367.

(53) *Ibid.*, p. 359.

(54) *Ibid.*, p. 370f.

(55) *Ibid.*, p. 386.

(56) *Ibid.*, p. 385.

(57) "The entry into Jerusalem was . . . Messianic for Jesus, but not for the people." *Ibid.*, p. 394.

(58) *Ibid.*, p. 307.

(59) Guenther Bornkamm, *Jesus of Nazareth* (New York: Harper & Row, 1960), p. 13.

(60) *Quest*, p. 398f.

(61) See N. Perrin, *What is Redaction Criticism?* (Philadelphia: Fortress, 1969).

(62) See Eric Graesser, *Das Problem der Parusieverzoegerung* (Berlin: Toepelmann, 1957). pp. 137-141.

(63) Quoted by Perrin, *Kingdom*, p. 32.

(64) *Ibid.*, p.33.

(65) *Ibid.*

(66) Cf. the comment of D. E. Nineham, "Schweitzer Revisited," in his *Explorations in Theology*, p. 129: "What Schweitzer has done is to show that when the historical method is applied to the New Testament, the result is not just to necessitate minor modifications in the picture it gives of Jesus, but to confront the timeless Christ of orthodoxy with a historical Jesus who, as such, inevitably belongs to a particular cultural and religious context and cannot belong to any other

time in the same immediate way that he belonged to his own." See also the insightful article of C. K. Barrett, "Albert Schweitzer and the New Testament," *ExTim* 87 (1975-76), 4-10.

(67) By picturing Jesus as an apocalyptic visionary, Schweitzer was not implying that Jesus was a deluded fanatic. In fact, Schweitzer expressly rejects such accounts of Jesus in his M.D. thesis, *The Psychiatric Study of Jesus*, trans. Charles R. Joy (Boston: Beacon Press, 1948).

(68) Hans Kueng, *On Being a Christian*, trans. Edward Quinn (London: Collins, 1974).

(69) "Q" stands for the "sayings" source used by Matthew and Luke. Why it was labelled "Q" is not certain -- see Neill, *Interpretation*, p. 119, n. 2.

(70) But see n. 44 above.

(71) See J. M. Robinson, *The Problem of History in Mark* (London: S.C.M. Press, 1957), esp. pp. 7-20.

(72) K. L. Schmidt, *Der Rahmen der Geschichte Jesu* (Berlin: Trowitsch & Sohn, 1919); for Wrede see n. 40 above.

(73) M. Dibelius, *Die Formgeschichte des Evangeliums* (Tuebingen: J. C. B. Mohr/Siebeck, 1919); ET: *From Tradition to Gospel*, trans. Bertram Lee Woolf (London: Ivor Nicholson and Watson Ltd., 1934).

(74) R. Bultmann, *Die Geschichte der synoptischen Tradition* (Goettingen: Vandenhoeck & Ruprecht, 1931); ET: *The History of the Synoptic Tradition*, trans. John Marsh (New York: Harper & Row, 1968).

(75) T. W. Manson, "The Life of Jesus: Some Tendencies in Present-Day Research," in *The Background of the New Testament and Its Eschatology*, ed. W. D. Davies and David Daube (London: Cambridge University Press, 1969).

(76) But see in this connection E. P. Sanders, *The Tendencies of the Synoptic Tradition* (London: Cambridge University Press, 1969).

(77) R. Bultmann, "The Primitive Christian Kerygma and the Historical Jesus," in *The Historical Jesus and the Kerygmatic Christ: Essays on the New Quest of the Historical Jesus*, ed. Carl E. Braaten and Roy A. Harrisville (Nashville: Abingdon, 1964), pp. 22-23.

(78) R. Bultmann, "Is Jesus Risen as Goethe?" in *Der Spiegel on the New Testament*, ed. W. Harenberg (New York: Macmillan, 1970), p. 233.

(79) See, for example, T. W. Manson, *The Teaching of Jesus: Studies of its Form and Content* 2nd ed. (Cambridge: Cambridge University Press 1935); V. Taylor, *The Formation of the Gospel*

Tradition (London: Macmillan, 1933); Harald Riesenfeld, *The Gospel Tradition and its Beginnings: A Study in the Limits of "Formgeschichte"* (London: Mowbray, 1957); B. Gerhardsson, *Memory and Manuscript: Oral Tradition and Written Transmission in Rabbinic Judaism and Early Christianity* (Lund: C.W.K. Gleerup, 1961), and his more recent *The Origins of the Gospel Traditions* (Philadelphia: Fortress, 1979).

(80) R. Bultmann, *Jesus and the Word* (New York: Charles Scribner's Sons, 1934, 1958).

(81) *Ibid.*, p. 4.

(82) *Ibid.*, p. 6.

(83) *Ibid.*, p. 11.

(84) R. Bultmann, "A Reply to the Theses of J. Schniewind," in *Kerygma and Myth: A Theological Debate*, ed. H. W. Bartsch (London: S.P.C.K., 1962), p. 117.

(85) R. Bultmann, "The Significance of the Historical Jesus for the Theology of Paul," in *Faith and Understanding I*, ed. Robert W. Funk, trans. Louise Pettibone Smith (New York: Harper & Row 1969), p. 241.

(86) "The work of existential philosophy, which I came to know through my discussions with Martin Heidegger, became of decisive significance for me. I found here the concept through which it became possible to speak adequately of human existence and therefore also of the existence of the believer." *The Theology of Rudolf Bultmann*, ed. Charles W. Kegley (New York: Harper & Row, 1966), p. xxiv.

(87) *Jesus and the Word*, p. 51.

(88) *Ibid.*

(89) Rudolf Bultmann, *Jesus Christ and Mythology* (New York: Charles Scribner's Sons, 1958), p. 61.

(90) *Ibid.*, p. 15.

(91) *Ibid.*, p. 61.

(92) Example cited by Ian Henderson, *Rudolf Bultmann* (London: Lutterworth Press, 1965), p. 41.

(93) Rudolf Bultmann, "The New Testament and Mythology," in *Kerygma and Myth*, p. 5.

(94) Karl Jaspers and R. Bultmann, *Myth and Christianity: An Inquiry into the Possibility of Religion without Myth* (New York: Noonday Press, 1958), p. 60.

(95) "New Testament and Mythology," p. 39.

(96) "New Testament and Mythology," pp. 1f.

(97) "Myth is an expression of man's conviction that the origin and purpose of the world in which he lives are to be sought not within it but beyond it -- that is, beyond the realm of known and tangible reality -- and that this realm is perpetually dominated and menaced by those mysterious powers which are its source and limit. Myth is also an expression of man's awareness that he is not lord of his own being. It expresses his sense of dependence not only within the visible world, but more especially on those forces which hold sway beyond the confines of the known. Finally, myth expresses man's belief that in this state of dependence he can be delivered from the forces within the visible world." *Ibid.*, pp. 10f.

(98) R. Bultmann, *Theology of the New Testament* 2 vols. (New York: Charles Scribner's Sons, 1951-55).

(99) Noted in the review of vol. 2 of *Theology of the New Testament* by J. M. Robinson in *Theology Today* 13 (1956-57), 261.

(100) *Jesus Christ and Mythology,* pp. 32f.

(101) This is actually the phrase of Bernard Lonergan -- see his *Method in Theology* (London: Darton, Longman, & Todd, 1972), p. 157.

(102) "Are we to read the Bible only as an historical document in order to reconstruct an epoch of past history for which the Bible serves as a 'source'? Or is it more than a source? I think our interest is really to hear what the Bible has to say of our actual present, to hear what is the truth about our life and about our soul." *Jesus Christ and Mythology,* p. 52.

(103) See "New Testament and Mythology," esp. pp. 27ff.

(104) Barth has an extended critique of Bultmann in *Kerygma and Myth,* vol. 2, "Rudolf Bultmann -- An attempt to understand him," pp. 83-132.

(105) So argues John Macquarrie, *The Scope of Demythologizing: Bultmann and His Critics* (New York: Harper, 1961).

(106) "New Testament and Mythology," pp. 43f.

(107) Ernst Kaesemann, "The Problem of the Historical Jesus" in his *Essays on New Testament Themes* (London: S.C.M. Press, 1964), pp. 15-47.

(108) "The Primitive Christian Kerygma and the Historical Jesus," p. 18.

(109) See esp. J. M. Robinson, *A New Quest of the Historical Jesus* (London: S.C.M. Press, 1959).

(110) Guenther Bornkamm, *Jesus of Nazareth,* trans. Irene and Fraser McLuskey with James Robinson (New York: Harper, 1960).

(111) In his book *Jesus and the Language of the Kingdom* (Philadelphia: Fortress, 1976), Norman Perrin argues that the Kingdom of God should not be understood as a *concept* but as a *symbol.* This is a most important book for those engaged in the interpretation of the Kingdom of God.

(112) Inauthentic existence, it will be recalled, is full of *Angst.* A clear description of what Bultmann means by inauthentic existence is found in his "New Testament and Mythology," pp. 18f.: "Since the visible and tangible sphere is essentially transitory, the man who bases his life on it becomes the prisoner and slave of corruption. An illustration of this may be seen in the way our attempts to secure visible security for ourselves bring us into collision with others; we can seek security for ourselves only at their expense. Thus on the one hand we get envy, anger, jealousy, and the like, and on the other compromise, bargainings, and adjustments of conflicting interests. This creates an all-pervasive atmosphere which controls all our judgements; we all pay homage to it and take it for granted. Thus man becomes the slave of anxiety (Rom 8:15). Everybody tries to hold fast to his own life and property, because he has a secret feeling that it is all slipping away from him."

(113) Authentic existence, it will be recalled, is to be freed from anxiety, and this can only come by faith in Jesus Christ: "The grace of God means *the forgiveness of sin,* and brings deliverance from the bondage of the past. The old quest for visible security, the hankering after tangible realities, and the clinging to transitory objects, is sin, for by it we shut out invisible reality from our lives and refuse God's future which comes to us as a gift. But once we open our hearts to the grace of God, our sins are forgiven; we are released from the past. This is what is meant by 'faith'; to open ourselves freely to the future. But at the same time faith involves obedience, for faith means turning our backs on self and abandoning all security. It means giving up every attempt to carve out a niche in life for ourselves, surrendering all our self-confidence, and resolving to trust in God alone, in the God who raises the dead (2 Cor 1:9) and who calls the things that are not into being (Rom 4:17). It means radical self-commitment to God in the expectation that everything will come from him and nothing from ourselves. Such a life spells deliverance from all worldly, tangible objects, leading to complete detachment from the world and thus to freedom." *Ibid.*, pp. 19f.

(114) There are basically three groups of scholars in this category, as has been pointed out by Etienne Trocmé, *Jesus and His*

Contemporaries (London: S.C.M. Press, 1973), pp. 2ff. "A first group believes it possible to fit all the data provided by our sources into the framework of an explanatory theory, usually quite bold, of the person and work of Jesus" (p. 2). See, for example, S. G. F. Brandon, *Jesus and the Zealots: A Study of the Political Factor in Primitive Christianity* (Manchester: Manchester University Press, 1967). A second group is the "traditionalists who are concerned above all to defend the historical truth of the whole gospel" (p. 4); e.g., H. Daniel-Rops, *Jesus in His Times* (London: Dutton, 1956). A third group attempts a "scholarly biography" of Jesus; e.g., V. Taylor, *The Life and Teaching of Jesus* (London: Macmillan, 1954).

(115) It is perhaps necessary to reiterate here that Strauss did *not* say Jesus himself was a mythical figure. Thus, Strauss's position is to be distinguished from the work of such as John Allegro, *The Sacred Mushroom and the Cross* (London: Hodder & Stoughton, 1970).

(116) See n. 67 above.

(117) See S. H. Mayor, "Jesus and the Christian Understanding of Society," *ScJTh* 32 (1979), 45-60.

(118) Indeed, this tendency is to be found in the New Testament itself, as a comparison of the Gospel of Mark with that of John shows.

(119) *Quest*, p. 403.

Notes to Chapter Two

(1) Cf. the comment of R. H. Lightfoot, *History and Interpretation in the Gospels* (London: Hodder & Stoughton, 1935), p. 225: "It seems as though the form of the earthly no less than the heavenly Jesus is for the most part hidden from us. For all the inestimable value of the Gospels, they yield to us little more than a whisper of his voice; we trace in them but the outskirts of his ways." This remark, incidentally, provoked some controversy, so much so that in a later publication Lightfoot felt obliged to clarify what he meant -- see his *The Gospel Message of St. Mark* (Oxford: Clarendon, 1950), p. 101, where he says he was echoing Job 26:24, the last words of which show "that the point of the passage lies in the contrast between the comparatively small knowledge which in Job's view is all that is available to man, and the boundless immensity which is quite beyond his grasp."

(2) *Theology of the New Testament*, I, p. 33.

(3) Cf. S. H. Mayor, "Jesus and the Understanding of Society," p. 48: "One of the functions of the Church is to interpret what otherwise might be mere abstractions, and make of them concrete guidance, something it does not by words only but by providing a context in which the experiment of Christian living might be attempted."

(4) Walter Bauer, *Orthodoxy and Heresy in Earliest Christianity*, ed. R. A. Kraft and G. Krodel (Philadelphia: Fortress, 1971). The original German version was *Rechtglaeubigkeit und Ketzerei im aeltesten Christentum* (Tuebingen: Mohr/Siebeck, 1934). A second edition, edited by Georg Strecker with the addition of two appendices, was published in 1963.

(5) There is now an extensive bibliography of works on orthodoxy and heresy in early Christianity. See esp. J. D. G. Dunn, *Unity and Diversity in the New Testament* (London: S.C.M. Press, 1977) and my own article "A Reflective Look at the Recent Debate on Orthodoxy and Heresy in Earliest Christianity," *Eglise et Théologie* 7 (1976), 367-378. J. M. Robinson, "Basic Shifts in German Theology," *Interpretation* 16 (1962), p. 77, describes Bauer's book as "in the older tradition of purely historical-critical scholarship." The debate which has ensued in the wake of Bauer's work has, however, often been polemical. Note, for example, the remark of E. Kaesemann, *The Testament of Jesus: A Study of the Gospel of John in the Light of Chapter 17* (London: S.C.M. Press, 1968), p. 75, n. 1. See also the

attack of J. Munck, "The New Testament and Gnosticism," *StTh* 15 (1961), 187, on Walter Schmithals.

(6) Bultmann himself endorsed Bauer's view (*Theology of the New Testament*, vol. 2, p. 137), and Bauer's thesis has been implicitly accepted by many Bultmannians. Thus, for example, E. Kaesemann posed the question "Does the New Testament Canon establish the unity of the Church?" and answered that rather it establishes "the plurality of confessions." ("The Canon of the New Testament and the Unity of the Church" in *Essays on New Testament Themes*, pp. 95-107). H. Koester, in acclaiming the work of Bauer, describes Christianity as "a religious movement which is syncretistic in appearance and conspicuously marked by diversification from the very beginning" ("GNOMAI DIAPHORAI: The Origin and Nature of Diversification in Early Christianity," *HarvThR* 58 [1965], 281).

(7) Bampton Lectures; London: Mowbray, 1954. Hereafter cited as *Pattern*.

(8) *Pattern*, p. 8.

(9) *Ibid*., p. 10.

(10) *Ibid*., p. 26.

(11) *Ibid*.

(12) *Ibid*., p. 27.

(13) *Ibid*., p. 28

(14) *Ibid*., p. 30, n. 2.

(15) *Ibid*., p. 30.

(16) *Ibid*., p. 31.

(17) *Ibid*.

(18) *Ibid*., p. 80.

(19 *Ibid*., p. 79.

(20) *Ibid*., pp. 102-142.

(21) Turner is aware of the problem of the relation of scripture to tradition: "The priority of Church to Bible is no doubt true in an instrumental sense. It was within the Christian Church and by her members that the New Testament was both written and received. Yet it is equally true and of even greater significance that the Scriptures, considered as the norm and ground work of the Church's life, remain prior to the Church. They enshrine the realities upon which her life must be based and to which they must be continually referred. There is, therefore, no formal contradiction between these two complementary propositions, and those who maintain a priority of value of Scripture over Tradition cannot on these grounds be accused of falling into an unobserved contradiction." *Ibid*., p. 487.

(22) *Ibid.*, p. 488.

(23) *Orthodoxy and Heresy,* pp. xxii-xxiii.

(24) As is evidenced by the misunderstanding of A. A. T. Ehrhardt, "Christianity before the Apostles' Creed," *HarvThR* 55 (1962), 73-119; reprinted in *The Framework of the New Testament Stories* (Manchester: Manchester University Press, 1964), pp. 151-199. Ehrhardt thinks that Bauer's work presupposes "that somewhere in Christianity a *regula fidei* was invented as a touchstone of orthodoxy, at the very outset of the history of the Church, an assumption which seems to leave out of consideration whether or not the problem of heresy was at all visualized in the early days of Christianity" (pp. 93, 172). This, of course, completely misconstrues Bauer's intention.

(25) *Orthodoxy and Heresy,* p. 314, n. 30.

(26) *Ibid.,*, pp. 270f., n. 84.

(27) *Ibid.*, p. 313, n. 29.

(28) *Ibid.*

(29) Bernard J. F. Lonergan, *Method in Theology,* p. 232: "History is not value-free in the sense that the historian refrains from all value-judgments. . . . The historian ascertains matters of fact, not by ignoring data, by failing to understand, by omitting judgments of value, but by doing all of these for the purpose of settling matters of fact."

(30) For an explanation of this term see *Method in Theology,* esp. pp. 81-89. See also Bruno Snell, *The Discovery of the Mind: The Greek Origins of European Thought* (Cambridge, Mass.: Harvard University Press, 1953). Snell's book describes how differentiations of consciousness took place in Greek thought.

(31) Lonergan, *Method in Theology,* p. 22: "Moreover, within method the use of heuristic devices is fundamental. They consist in designating and naming the intended unknown, in setting down at once all that can be affirmed about it, and in using this explicit knowledge as a guide, a criterion, and/or a premise in the effort to arrive at fuller knowledge."

(32) *Orthodoxy and Heresy,* p. 132. However, Bauer does at several places indicate the undifferentiated and fluid character of early Christian thought. Note especially: "The two types of Christianity were not at all clearly differentiated from each other . . ." (p. 59); "Moreover, in this early period 'orthodoxy' is just as much a sort of collective concept as is 'heresy', and clothes itself in quite different forms according to the circumstances . . ." (p.77); "The confession of

Jesus as Lord and heavenly redeemer is a common foundation for both tendencies, and for a long time sufficed to hold the differently oriented spirits together in *one* fellowship . . ." (p. 90).

(33) "GNOMAI DIAPHORAI," p. 281.

(34) "Christianity before the Apostles' Creed," pp. 93f. It is ironic that, as we have noted, Ehrhardt himself had missed the point of Bauer's book.

(35) It is perhaps instructive to quote the total context of Koester's remark: "On the other hand, the search for theological criteria cannot be avoided by means of a retreat into dogmatic or religious propositions. Such propositions often attempt to fill the gaps and bridge the inconsistencies in the history of orthodoxy by postulating a primitive orthodox Church which concealed its true beliefs in certain practices and institutions, and in the -- theologically mute -- 'lex orandi' " ("GNOMAI DIAPHORAI," p. 281).

(36) Lonergan, *Method in Theology,* p. 236: "[H]orizons may differ genetically. They are related in successive stages in some process of development. Each later usage presupposes earlier stages, partly to include them and partly to transform them. Precisely because the stages are earlier and later, no two are simultaneous. They are parts, not of a single communal world, but of a single biography or of a single history."

(37) *Pattern,* p. 474.

(38) *Pattern,* p. 477.

(39) It is interesting to note that the differences were almost always over how the original deposit of faith could best be safeguarded, for both the heretics and the orthodox claimed to do this.

(40) "Thus orthodoxy in its totality proved to be a far richer thing than any of its components . . . in each case it was the tradition defeated in the sphere of formulation which was ultimate in victory in the realm of interpretation." *Pattern,* p. 476.

(41) *Pattern,* p. 498.

(42) The quest for balance is often lacking in contemporary debate, e.g., D. Welbourn, *God-Dimensional Man* (London: S.C.M. Press, 1972) attempts to formulate a Christology based on Jesus solely as a human being. See in contrast G. V. Jones, *Christology and Myth in the New Testament* (London: Allen & Unwin, 1956), who stresses that Christology must balance the anagogic and catagogic aspects of Christ. See also the Christological debates in *Christ, Faith and History,* ed. S. W. Sykes and J. P. Clayton (Cambridge: Cambridge University Press, 1972). Here M. F. Wiles's essay, "Does Christology rest on a mistake?" pp. 3-12, reprinted from *Religious*

Studies 6 (1970), 69-76, is particularly important. See also his "In defence of Arius," *JTS* n.s. 13 (1962), 339-347. Also *Myth of God Incarnate*, ed. John Hick (London: S.C.M. Press, 1977) and *Incarnation and Myth: The Debate Continued*, ed. Michael Goulder (London: S.C.M. Press, 1979).

(43) *The Apostolic Preaching and Its Developments* (London: Hodder & Stoughton, 1936), pp. 73f.

(44) Turner, *Pattern*, p.488, defines coherence as a hypothesis which is "self-consistent and congruous with other related fields of knowledge and experience." Cf. J. H. Newman, *Oxford University Sermons* (London: Rivingtons, 1892), p. 337: "[I]t being a definition of heresy, that it fastens on some one statement as if the whole truth, to the denial of all others, and as the basis for a new faith."

(45) It is salutary to recall here that it was Newman who pointed out that if there was a phenomenon named Christianity which the historian could investigate, Christianity was a fact. As such a fact, it "interfered with" such theories as, for example, "historically it [Christianity] has no substance of its own but [has been] a mere assemblage of doctrines and practices derived from without" -- *An Essay on the Development of Christian Doctrine* (1878; repr. London: Sheed and Ward, 1960), section 2 of the Introduction. Note also the following remark of Newman: "The phenomenon, admitted on all hands, is this: That a great portion of what is generally received as Christian truth is, in its rudiments or in its separate parts, to be found in heathen philosophies and religions. For instance, the doctrine of a Trinity is to be found both in the East and in the West; so is the ceremony of washing; so is the rite of sacrifice. The doctrine of the Divine Word is Platonic; the doctrine of the Incarnation is Indian; of a divine kingdom is Judaic; of Angels and demons is Magian; the connection of sin with the body is Gnostic; celibacy is known to Bonze and Talapoin; a sacerdotal order is Egyptian; the idea of a new birth is Chinese and Eluesinian; belief in sacramental virtue is Pythagorean; and honours to the dead are a polytheism. Such is the general nature of fact before us, Mr. Milman argues from it -- 'These things are in heathenism, therefore they are not Christian': we, on the contrary, prefer to say, 'These thing are in Christianity, therefore they are not heathen.' " *Ibid.*, viii, 2 (12). This does indicate a fundamental difference in attitude which is to some extent still present in the contemporary debate.

(46) Lonergan, *Method in Theology*, p. 138: "The unity of a subject in process of development is dynamic. For as long as further advance is possible, the perfection of complete immobility has not been attained, and, for that reason, there cannot yet be reached the

logical ideal of fixed terms, accurately and immutably formulated axioms, and absolutely rigorous deduction of all possible conclusions. The absence, however, of static unity does not preclude the presence of dynamic unity. . . ."

Note also G. Biemer, *Newman on Tradition* (London: Burns & Oates, 1967), p. 129: "The static and dynamic elements of revelation are, by their nature, and by reason of the medium of time, essential to the conservation and life of the deposit. To abandon either of them in favour of the other is to disregard history in one way or another. Either the origins are neglected, to make way for 'enthusiasm' or the fanciful, or the continuity of history is not taken seriously, and an anachronistic 'classicism' holds the field. Newman combined both elements in his theory of development." This seems very close to Turner's own attempt.

(47) In a private communication to me, Professor Turner pointed out that his view of the importance of the *lex orandi* is confirmed not only by the evidence which he adduces for basic and theologically unreflective statements both of the Divinity of Christ and the Trinity but also from the Church's reaction to heresy. This is neither a position of complete closedness as the classicist theory presupposed nor a complete openness as Bauer and others implied. The reaction to Gnosticism is a case in point. We may grant that orthodoxy and heresy existed side by side in some areas and that there was a penumbra between them. However, there was sufficient *communis sensus fidelium* to suspect something amiss in Gnosticism and to put the Church on its guard and at "action stations" against the new movement. "Action stations" expressed themselves initially in the crystallization of existing or developing institutions such as episcopacy, the Canon of Scripture, and the Creeds. Moreover, Irenaeus juxtaposed the Gnostic systems and the basic Christian *credenda* and allowed the former to fall by their own weight. Clement tries to show that the Christian is the true Gnostic. Origen offers a viable alternative based on the rule of faith but cuts the ground from under the feet of his opponents by offering a more intelligible and coherent alternative.

(48) B. F. Meyer, *The Church in Three Tenses* (Garden City, New York: Doubleday & Co., 1971), p. 71.

(49) It must be said here that Christian self-understanding in every age, although always involved in development, always shared Turner's presupposition and never Koester's. This surely weakens the claim of Koester and others that they have caught hold of the earliest series of Christian mentalities. It may be, on the basis of historical investigation alone, that we cannot show that Christianity has had a permanent identity maintained through extraordinary cultural

mutations and maintained, in part, through intolerance for heresy; but we *can* say on the basis of historical investigation alone that this is the account that Christianity has always given of itself.

(50) By this phrase Turner means both the reality of God and the gift God makes of himself in his Son.

(51) See B. F. Meyer, *Church in Three Tenses*, p. 65.

(52) And thus can hardly be "theologically mute"! On the contrary, it is that from which all authentic theological discourse flows.

(53) It is a hermeneutical principle that there is a "circle of things and words" (*Sache und Sprache*). That is, I understand words by understanding the things they refer to; I understand things by understanding the words that refer to them. The first limb states the more fundamental insight -- it explains why a blind man will find a lecture on colour obscure.

(54) Cf. G. Biemer, *Newman on Tradition*, p. 122: "Having drawn the necessary consequences and formed the necessary definitions, reason arrives at a whole theological system. But its activity does not end there. The urge to know pushes it still further, and the growth of the spirit can never be checked. The end could only be brought about by the object, if all its aspects were grasped in their totality, and the idea then fully comprised and comprehended in an adequate act of knowledge. But such a perfection of knowledge cannot be attained even with regard to the objects and systems of our own world and its reality. Even here there are always new possibilities to explore, new aspects to throw light on, new means for approximating to the essence of the thing itself. Much less is it possible to have anything like complete comprehension of revelation, where human reason is confronted with the mysterious truths of the Infinite."

(55) B. F. Meyer, *Church in Three Tenses*, p. 140.

(56) J. Denney, *Jesus and the Gospel: Christianity Justified in the Mind of Christ* (London: Hodder & Stoughton, 1919), p. 101.

(57) The use of this metaphor argues for the identity of doctrine throughout the successive stages of Christianity. Meyer, *Church in Three Tenses*, pp. 75f., objects to the use of the analogy of organic growth because, he feels, it implies that the Church is predestined to "a fixed scheme of development through infancy, youth, and maturity to decline, senescence, and death." The use of the analogy of organic growth is useful, however, in excluding transformism.

(58) The important works in this debate are: R. Sohm, *Kirchenrecht*, I (Leipzig: Dunker & Humblet, 1892); A. Harnack, *Entstehung und Entwicklung der Kirchenfassung und des Kirchenrechts in den*

zwei ersten Jahrhunderten (Leipzig: J. C. Hinrichs' Buchhandlung, 1910); O. Linton, *Das Problem der Urkirche in der neueren Forschung: eine kritische Darstellung* (Uppsala: A'lmqvist & Wiksells boktryckeri.-a.-b., 1932). A brief summary of the issues appears in H. Conzelmann, *An Outline of the Theology of the New Testament* (London: S.C.M. Press, 1969), pp. 41f. See also K. L. Schmidt's article on *ekklesia* in *TDNT*; A. M. Hunter, *The Unity of the New Testament* (London: S.C.M. Press, 1943), ch. 6.

(59) See L. Cerfaux, *The Church in the Theology of St. Paul* (New York: Herder, 1963), pp. 248-261.

(60) In this connection see the interesting article by N. A. Dahl, "The Atonement--An Adequate Reward for the Akedah? (Ro. 8:32)," in *Neotestamentica et Semitica. Studies in Honour of Matthew Black*, ed. E. Earle Ellis and Max Wilcox (Edinburgh: T. & T. Clark, 1969), pp. 15-29.

(61) It is true that the law is viewed negatively by Paul. But nevertheless for Paul the gospel message is a fulfilment of the salvific hope to which the law pointed, even if in a negative way.

(62) J. N. D. Kelly, *Early Christian Creeds* (London: Longmans, 1960), p. 8.

(63) See n. 43 above. However, it should be noted here that Dodd's thesis has been subjected both to attack and modification. The key question is: Are the kerygmatic speeches isolated by Dodd free compositions from the hand of Luke? Perhaps the most important article advocating the free composition theory is that of C. F. Evans, "The Kerygma," *JTS*, n.s. 7 (1956), 25-41. His central argument is that there is no *Sitz im Leben* for the repetition and preservation of the speeches of the Apostles. That is, he sees no purpose corresponding to the interests of the primitive Church which would motivate the recording of the early kerygmatic discourses. See also the criticisms of Dodd's argument by H. J. Cadbury, "Acts and Eschatology," in *The Background to the New Testament and its Eschatology. Studies in Honour of C. H. Dodd*, ed. W. D. Davies and D. Daube (Cambridge, Mass.: Cambridge University Press, 1956), pp. 300-321, esp. pp. 313ff.

On the other hand, there are those who seek to isolate Aramaic substrata in the kerygmatic speeches of Acts, thus indicating a primitive source. Here there are various works, each of which formulates a slightly different hypothesis; e.g., C. C. Torrey, *The Composition and Date of Acts* (Cambridge, Mass.: Harvard University Press, 1916); W. L. Knox, *The Acts of the Apostles* (Cambridge: Cambridge University Press, 1948), pp. 18-21; M. Black, *An Aramaic Approach to the*

Gospels and Acts (Oxford: Oxford University Press, 1967); M. Wilcox, *The Semitisms of Acts* (Oxford: Clarendon, 1965).

However, the argument that the kerygmatic speeches in Acts do reflect a primitive Palestinian *kerygma* does not stand or fall with Aramaic source theories. There is another -- theological -- criterion: theological archaisms. For example, in Acts the death and resurrection of Jesus are dissociated inasmuch as the death of Jesus is the work of evil men, while the resurrection is the supreme saving act of God. In the Synoptics, Paul and John, the death and resurrection are grasped as a unity as the saving event. See D. M. Stanley, "The Conception of Salvation in Primitive Christian Preaching," *CBQ* 18 (1956), 231-254. See also E. Schweizer, "Zu den Reden der Apostelgeschichten," *TZ* 13 (1957), 1-11, who seeks to demonstrate the use of ancient material in the specifically Christological parts of the speeches.

In conclusion, we should perhaps say that although the view that the speeches of Acts do reflect a primitive Palestinian *kerygma* has been the subject of attack by many scholars, the issue has certainly not been resolved. In any case, our view does not stand or fall with this question, for we place greater emphasis on the early faith formulas.

(64) *Early Christian Creeds,* p. 23.

(65) *Paralambein* and *paradidonai* are fixed concepts of tradition. In all probability they correspond to *qibbel* and *masar* (so O. Cullmann, *The Early Church* [London: S.C.M. Press, 1956], p. 63, against E. Norden, *Agnostos Theos. Untersuchungen zur Formgeschichte religioeser Rede* [Stuttgart: Teubner, 1913], p. 270). In present day vernacular "tradition" denotes an unwritten body of doctrine handed down through the Church, but originally the word emphasized authoritative delivery. Hence in the early Church the word usually connoted that doctrine (written or unwritten) which Jesus or his Apostles committed to the Church -- see J. N. D. Kelly, *Early Christian Doctrines* (London: A. & C. Black, 1960), pp. 30f.

(66) J. Jeremias, *Die Abendmahlsworte Jesu* (Goettingen: Vandenhoeck & Ruprecht, 1960), pp. 95-97; *ibid.*, "Artikelloses Christos. Zur Ursprache von I Kor. XV 3b-5," *ZNW* 57 (1966), 211-215; also B. Klappert, "Zur Frage des semitischen der greichischen Urtextes von I Kor. XV 3-5," *NTS* 13 (1967), 168-173, against H. Conzelmann, "Zur Analyse der Bekenntnisformel I Kor. 15, 3-5," *EvTh* 25 (1965), 1-11.

(67) See B. F. Meyer, *The Aims of Jesus* (London: S.C.M. Press, 1979), pp. 61-63.

(68) Walter Eichrodt, *Theology of the Old Testament,* vol. 1 (London: S.C.M. Press, 1961), pp. 32f.

(69) John Macquarrie, *Paths in Spirituality* (New York: Harper & Row, 1972), p. 11.

(70) *Ibid.*, p. 54.

(71) *Ibid.*, p. 58.

Notes to Conclusion to Part I

(1) A good example of a "biblicist" approach is found in the following newspaper article:

> "Interview" with Jesus Christ
>
> Vatican City (AP) -- "The Vatican radio presents an interview with Jesus," news commentator Paolo Scappucci announced.
>
> Scappucci said he had arranged a 15-minute conversation with Jesus Christ, "who will speak through an eminent biblical scholar" on pressing contemporary issues . . .
>
> Here are some questions and answers from the programme:
>
> Q: What would you say to young heroin addict?
> A: Every one who drinks of this water will thirst again. He, however, who drinks of the water that I will give him shall never thirst. (John 4:13)
> Q: And to the fanatics who carry pistols and Molotov cocktails?
> A: For all those who take the sword will perish by the sword. (Matthew 26:52)
> Q: What's the greatest risk facing democracy today?
> A: Every Kingdom divided against itself is brought to desolation, and every city or house divided against itself will not stand. (Matthew 12:25)
> Q: What do you think of social conflict? Is the use of violence to fight the establishment justified?
> A: Blessed are the peacemakers, for they shall be called children of God. (Matthew 5:9)
> Q: How do you see the world today?
> A: I have not come to judge the world but to save the world. (John 12:47)

(Taken from *The Western Star* [Corner Brook, Newfoundland], Saturday, May 19, 1979.)

Here the principle of Jesus-as-the-norm for Christian belief and practice is applied in an apparently easy and direct way. You ask a question and search for words of Jesus which will supply an answer.

There are a number of drawbacks to this approach. In the first place, as we have seen in chapter one, the historicity of many of Jesus' sayings is disputed. Secondly, this kind of biblicist approach is selective of Jesus' answer. The answer to the second question, for example, could just as easily be "Let him who has no sword sell his mantle and buy one" (Lk 22:36). The answer to the fourth question might have been "Do not think that I have come to bring peace on earth; I have not come to bring peace, but a sword. For I have come to set a man against his father, and a daughter against her mother, and a daughter-in-law against her mother-in-law; and a man's foes will be those of his own household" (Mt 10:34-36). Verses can be snatched from scripture to justify almost every standpoint, even diametrically opposed ones. But an even more serious criticism of the biblicist approach is that it completely misconstrues the nature of the Bible. The Bible is not a book of instructions. It does not give *directions* for one's life, rather *a direction* -- see H. Evan Runner, *Scriptural Religion and Political Task* (Toronto: Wedge Publishing Foundation, 1974).

(2) *Out of My life and Thought*, p. 56.

(3) *Ibid.*, p. 55f.

(4) See *Quest*, p. 399.

(5) *Out of My Life and Thought*, p. 56.

(6) *Jesus and the Word*, p. 97. It is interesting to read Mk 10:17-22 in the light of this comment.

(7) *Ibid.*, p. 109.

(8) H. E. W. Turner, "Orthodoxy and the Church Today," *The Churchman* 86 (1972), p. 167.

(9) J. A. T. Robinson, *Honest to God* (London: S.C.M. Press, 1963), pp. 112f.

(10) *On Being a Christian*, p. 544.

(11) *Ibid.*, p. 545.

(12) There is a distinct difference between following Jesus and imitating him. Jesus pursued a lifestyle which, if imitated quite literally, would lead to the collapse of Western society overnight. We have unfortunately so domesticated Jesus that we fail to see him as a person who repudiated possessions and family life (he remained unmarried and severed himself from his family) and had no regular job. Moreover, Jesus made demands which fly in the face of all social conventions. Take, for example, the following two sayings of Jesus: "Leave the dead to bury their own dead" (Lk 9:60a) and "If anyone strikes you on the right cheek, turn to him the other also; and if anyone would use you and take your coat, let him have your cloak as

well; and if anyone forces you to go one mile, go with him two miles" (Mt 5:39b-41). As Norman Perrin, *Jesus and the Language of the Kingdom*, p. 51, comments, "To 'leave the dead to bury their own dead' is to act so irresponsibly as to deny the very fabric which makes possible communal existence in the world; it is a fundamental denial of the kind of personal and communal sense of responsibility which makes possible the act of living in community. The giving of the 'cloak as well' and the going of the 'second mile' are commandments, and it is impossible to take them literally as moral imperatives. In the first one, the results would be 'indecent exposure' and in the second a lifetime of impressed service."

(13) C. K. Barrett, *Jesus and the Gospel Tradition* (London: S.P.C.K., 1967), p. 107. Barrett is quoting from Luther's lecture on Psalm 5. See Luther, *Lecture on Psalm 5*, Weimar edition, *Luther's Works*, V, p. 183: "One becomes a theologian by living, by dying, and by being damned -- not by understanding, reading, and speculating." Barrett is suggesting that for "theologian" Luther might have put "Christian."

Notes to Chapter Three

(1) C. S. Lewis, "De Descriptione Temporum" in *They Asked for a Paper* (London: Geoffrey Bles, 1962), pp. 9-25.

(2) A. Koestler, *The Ghost in the Machine* (London: Hutchinson, 1967).

(3) Sir Herbert Butterfield, *The Origins of Modern Science, 1300-1800* (New York: Free Press, 1957) is still the best book on the subject. Butterfield's time reference is, however, wider than 1540-1700. See also A. E. E. McKenzie, *The Major Achievements of Science*, Vol. I (Cambridge: Cambridge University Press, 1967). On the early Greek view of science, see G. E. R. Lloyd, *Early Greek Science: Thales to Aristotle* (London: Chatto & Windus, 1970).

(4) *The Ghost in the Machine*, p. 301f.

(5) Aristotle's astronomical ideas had been modified by the second century A.D. astronomer Ptolemy. It was the Ptolemaic system -- which still had the earth as the centre -- which was universally adhered to in the Middle Ages.

(6) Quoted in P. Miller and K. Pound, *Creeds and Controversies* (London: English Universities Press Ltd., 1969), p. 184.

(7) F. Bacon, *The Advancement of Learning*, Book I, ed. H. G. Dick (New York: Random House, 1955), p. 193.

(8) Darwin quite freely acknowledged his debt to others -- see the section entitled "An Historical Sketch of the Progress of Opinion on the Origin of Species, previously to the Publication of this Work" in the third edition of *On the Origin of Species by Means of Natural Selection, or the Preservation of Favoured Races in the Struggle for Life* (London: John Murray, 1861).

(9) As, for example, in so-called Social Darwinism. Herbert Spencer (1820-1903) attempted to apply Darwin's theory to change and development in human societies. He believed that Darwin's theory implied that some people are more "naturally" capable than others and that these people become the wealthy and influential. Hence, Spencer writes: "The poverty of the incapable, the distresses that come upon the imprudent, the starvation of the idle, and those shoulderings aside of the weak by the strong . . . are the decrees of a large, farseeing benevolence" (cited in Sidney Fine, *Laissez Faire and the General Welfare State* [Ann Arbor: University of Michigan Press, 1964], p. 38).

(10) Langdon Gilkey, *Naming the Whirlwind: The Renewal of God-Language* (Indianapolis: Bobbs-Merrill Co., 1969), p. 40, n. 1.

(11) See C. S. Carter, *A Hundred Years of Evolution* (London: Sidgwick & Jackson, 1957), p. 22.

(12) *Vestiges of the Natural History of Creation* (London: J. Churchill, 1844), p. 232.

(13) *The Autobiography of Charles Darwin*, ed. Nora Barlow (London: Collins, 1958), p. 87.

(14) *The Life and Letters of Charles Darwin Including an Autobiographical Chapter*, ed. F. Darwin, Vol. I (New York, 1898), p. 83. Cited in James C. Livingston, *Modern Christian Thought: From the Enlightenment to Vatican II* (New York: Macmillan, 1971), p. 226.

(15) London: John Murray, 1859.

(16) Darwin wrote to Alfred Wallace in 1857: "You ask me whether I shall discuss 'man'. I think I shall avoid the whole subject, as so surrounded with prejudices; though I fully admit that it is the highest and most interesting problem for the naturalist." Cited in John C. Greene, *The Death of Adam* (Ames, Iowa: The Iowa State University Press, 1959), p. 309.

(17) "Darwin's *Origin of Species*," *Quarterly Review* 108 (1860), 225-264. Wilberforce's authorship of this article was not made public until it was reprinted in his *Essays Contributed to the "Quarterly Review"* 2 vols. (London, 1874).

(18) Not all clergymen totally rejected Darwin's theory -- see Livingston, pp. 231ff.

(19) Reissued in New York by P. F. Collier & Son, 1902.

(20) *Ibid.*, p. 528.

(21) Livingston, p. 228.

(22) Philip Rieff, *Freud: The Mind of the Moralist* (New York: Doubleday, 1959), p. ix, says, "Freud created the masterwork of the century, a psychology that connects fathers and mothers, lovers and haters, sick and less sick, the arts and the sciences, that unriddles, to use Emerson's prophetic catalogue of subjects considered inexplicable in his day -- 'language, sleep, madness, dreams, beasts, sex.' Freud's doctrine, created piecemeal and fortunately never integrated into one systematic statement, has changed the course of Western intellectual history, moreover, it has contributed as much as a doctrine can to the correction of our standards of conduct."

(23) Trans. & ed. James Strachey (New York: Basic Books, 1958).

(24) Clyde Kluckholn, "Recurrent Themes in Mythology," *The Making of Myth*, ed. Richard Malin Ohmann (New York: Putnam, 1962), pp. 53-63, showed that the Oedipus Complex was not as prevalent in mythology as Freud's theory seemed to require. Most of Freud's own disciples (e.g., Adler, Jung, Fromm, Karen Horney) have disputed the universality of the Oedipus Complex.

(25) In *Sociology and Religion: A Book of Readings*, ed. Norman Birnbaum and G. Lenzer (Englewood Cliffs: Prentice-Hall, 1969), pp. 168-173. The original article, "Zwanshandlungen und Religionsuebung," was written in 1907 and is to be found in Freud's *Gesammelte Shriften* 10 (1924-34), pp. 210ff.

(26) *Ibid.*, p. 173.

(27) Trans. A. A. Brill (New York: Vintage Books, 1918).

(28) *Ibid.*, p. 185.

(29) See A. C. Kroeber, "Totem and Taboo: An Ethnologic Analysis," in *A Reader in Comparative Religion*, ed. W. A. Lessa and E. Z. Vogt (New York: Harper & Row, 1965), pp. 45-53. Freud was unmoved by his critics -- see his *Moses and Monotheism*, trans. Katherine Jones (New York: Vintage, 1939), p. 169.

(30) Trans. W. D. Robson-Scott (New York: Doubleday, 1961).

(31) *Ibid.*, p. 71.

(32) S. Freud, *New Introductory Lectures on Psychoanalysis*, trans. W. J. H. Sprott (London: Hogarth Press, 1949), pp. 214ff.

(33) Alasdair MacIntyre, *Difficulties in Christian Belief* (London: S.C.M. Press, 1959), p. 95.

(34) *The Difference Between the Democritean and Epicurean Philosophy of Nature*, trans. N. Livergood (The Hague: Martinus Nijhoff, 1967), p. 62.

(35) *Ibid.*, p. 61.

(36) "Contribution to the Critique of Hegel's 'Philosophy of Right': Introduction," *Early Writings*, trans. T. B. Bottomore (New York: McGraw-Hill, 1964), p. 43.

(37) "Paris Manuscripts," *Early Writings*, p. 214.

(38) In this Marx was appropriating Feuerbach's criticism of Hegel -- see Kathleen L. Clarkson and David J. Hawkin, "Marx on Religion: The Influence of Bruno Bauer and Ludwig Feuerbach on His Thought and its Implications for the Christian-Marxist Dialogue," *ScJTh* 31 (1978), pp. 543ff.

(39) Karl Marx, *Capital*, ed. F. Engels, trans. Samuel Moore and Edwood Avelling (New York: The Modern Library, 1906), pp. 83, 91.

(40) Karl Marx & F. Engels, *The German Ideology,* trans. S. Ryazanskaya (Moscow: Progress Publishers, 1964), pp. 60-61.

(41) "Introduction," p. 43.

(42) *The Holy Family,* trans. R. Dixon (Moscow: Foreign Languages Publishing House, 1956), p. 171.

(43) "Paris Writings," p. 156.

(44) *Ibid.*, p. 166f.

(45) See "Excerpt-Notes of 1844" in *Writings of the Young Marx on Philosophy and Society,* trans. Lloyd Easton and Kurt Guddat (New York: Doubleday, 1967), pp. 271ff.

(46) This is well argued by Jacques Ellul -- see below, chapter 4.

(47) See esp. Leo Strauss, "The Three Waves of Modernity," in *Political Philosophy: Six Essays by Leo Strauss*, ed. with introduction by Hilail Gildin (New York & Indianapolis: Pegasus, 1975), pp. 81-98.

(48) *Ibid.*, p. 84.

(49) Machiavelli (1469-1527), of course, predates the Scientific Revolution.

(50) J. H. Whitfield, *Discourses on Machiavelli* (Cambridge: Heffer, 1969).

(51) In defence of this "old-fashioned and simple opinion" see Leo Strauss, *Thoughts on Machiavelli* (Seattle & London: University of Washington Press, 1958), esp. the Introduction. It is interesting that Quentin Skinner, in his recent *Machiavelli* (Cambridge: Cambridge University Press, 1981), demurs from value judgments on Machiavelli's thought.

(52) Francis Bacon, *The Advancement of Learning,* ed. Arthur Johnston (Oxford: Oxford University Press, 1974), p. 157.

(53) See Strauss, *Thoughts on Machiavelli,* pp. 15-53.

(54) *The Prince,* ch. 15. The translation is from Robert M. Adams, *The Prince* (New York: W. W. Norton & Co. Inc., 1977), p. 44. All subsequent quotations are taken from this edition.

(55) See, for example, *Discourses,* Book I, ch. 3. Machiavelli's low opinion of human nature receives pointed expression in his comedy *Mandragola.*

(56) Warren Winiarski, "Niccolò Machiavelli," in *History of Political Philosophy,* ed. Leo Strauss and Joseph Cropsey (Chicago: Rand McNally, 1963), p. 273.

(57) Strauss, "Three Waves of Modernity," p. 87.

(58) Leon R. Kass, "The New Biology: What Price Relieving Man's Estate?" in *Science, Technology and Freedom*, ed. Willis H. Truitt and T. W. Graham Solomons (Boston: Houghton Mifflin Co., 1974), p. 165.

(59) F. Nietzsche, *The Use and Abuse of History*, trans. Adrian Collins (Indianapolis: Bobbs-Merrill, 1957), p. 61.

(60) *The Joyful Wisdom*, trans. Walter Kaufmann, *The Portable Nietzsche* (New York: Viking Press, 1968), pp. 95f.

(61) George Grant, *Time as History* (Toronto: Canadian Broadcasting Corporation, 1969), p. 30.

(62) *Thus Spake Zarathustra* in *The Portable Nietzsche*, pp. 129ff.

(63) We have made no mention, for example, of such key ideas as the *Uebermensch*, *amor fati*, and the eternal return.

(64) This is argued by Jacques Ellul -- see ch. 4.

(65) See, for example, Vance Packard, *The People Shapers* (London: Bantam, 1979), and Leon Kass, "The New Biology."

(66) G. P. Grant, *Technology and Empire* (Toronto: House of Anansi, 1969), p. 114, n. 3.

(67) Gad Horowitz, "Red Tory" in *Canada: A Guide to the Peaceable Kingdom*, ed. William Kilbourn (Toronto: Macmillan, 1970), p. 257.

(68) *Ibid.*

(69) "The New Biology," p. 164.

Notes to Chapter Four

(1) Trans. Talcott Parsons (New York: Scribners, 1958). For a somewhat different approach see R. H. Tawney, *Religion and the Rise of Capitalism: A Historical Study* (New York and Toronto: Mentor, 1954), and especially his criticism of Weber in ch. 4, n. 32.

(2) See, e.g., Leo Strauss, *Natural Right and History* (Chicago: University of Chicago Press, 1953), p. 60, n. 22, where he argues that Puritanism was more of a "carrier" of the "new philosophy" created by the teachings of Machiavelli, Bacon, and Hobbes.

(3) F. Gogarten, *The Reality of Faith,* trans. Carl Michalson *et al.* (Philadelphia: Westminster, 1959), p. 187.

(4) See Larry Shiner, *The Secularization of History: An Introduction to the Theology of Friedrich Gogarten* (New York: Abingdon, 1966), p. 28. This book is an excellent introduction to the difficult thought of Gogarten. In much of what follows I am indebted to Shiner.

(5) Shiner, p. 51.

(6) F. Gogarten, *Despair and Hope for our Time,* trans. Thomas Wieser (Philadelphia: Pilgrim Press, 1970), p. 87.

(7) *Reality of Faith,* p. 167.

(8) S. Paul Schilling, *Contemporary Continental Theologians* (London: S.C.M. Press, 1966), p. 105.

(9) This is Shiner's rendition of Gogarten's phrase "Gottes Zukuenftigkeit." See Shiner, p. 74.

(10) *Despair and Hope for our Time,* p. 135.

(11) *Contemporary Continental Theologians,* p. 112.

(12) Schilling thinks that Gogarten pushes the distinction between faith and works too far, consigning them to separate compartments with no relation to each other, *ibid.,* 117ff.

(13) Harvey Cox, *The Secular City: Secularization and Urbanization in Theological Perspective* (New York: Macmillan, 1965). Two other books by Cox of interest to our inquiry are *On Not Leaving it to the Snake* (London: S.C.M. Press, 1968) and *God's Revolution and Man's Responsibility* (Valley Forge: Judson Press, 1965).

(14) "Secularity" is used to connote a this-worldly, affirmative, and responsible attitude towards history and society. It is thus to be distinguished from "secularism," which is an anti-religious philosophy

derived from a secular mode of existence. In *Secular City* Cox (following Gogarten) distinguishes between "secularization" (which is harmonious with biblical faith) and "secularism" (pp. 18-21).

(15) H. Cox, "The Christian in a World of Technology" in *Science and Religion; New Perspectives on the Dialogue*, ed. Ian G. Barbour (New York and Evanston: Harper & Row, 1968), p. 262.

(16) It should, however, be noted that the implications of this acceptance are often discerned differently. The so-called radical theologians of the 1960s argued that biblical faith had been rendered irrelevant by secularity. But such as Cox argue that biblical faith is the basis of secular society. We should perhaps also note in passing that some of the radical theologians have modified the stand they took in the sixties -- see Paul van Buren, *Discerning the Way: A Theology of the Jewish-Christian Reality* (New York: Seabury, 1980) and Thomas J. J. Altizer, *The Self-Embodiment of God* (San Francisco: Harper and Row, 1977).

(17) The phrase was originally Emmanuel Kant's.

(18) "Last Letters from a Nazi Prison" in *The New Christianity*, ed. W. R. Miller (New York: Dell, 1967), p. 280.

(19) *Ibid.*, 284.

(20) *Letters and Papers from Prison* (London: S.C.M. Press, 1951). This book was issued in the U.S.A. under the title *Prisoner of God* (New York: Macmillan, 1954). There is some dispute about how well the thought of Bonhoeffer's earlier writings accords with his later ones -- see David H. Hopper, *A Dissent on Bonhoeffer* (Philadelphia: Westminster, 1975), esp. ch. 2.

(21) See esp. *Thus Spake Zarathustra*, pp. 98-105.

(22) Edmund Leach, *A Runaway World?* (New York: Oxford University Press, 1968).

(23) Cf. Rabbi Richard L. Rubenstein, *After Auschwitz: Radical Theology and Contemporary Judaism* (Indianapolis and New York: Bobbs-Merrill, 1966), p. 254: "It is only when we regard human beings as replaceable ciphers whose role is to keep the machinery of the technopolis functioning that the tragic sense is lost. The loss of the tragic dimension does not lead to a new optimism but to the depersonalization and dehumanization of life and death alike."

(24) Ellul's lack of professional theological training does not mean that theologians can afford to ignore him -- see R. R. Ray, "Jacques Ellul's Innocent Notes on Hermeneutics," *Interpretation* 33 (1979), 268-282.

(25) See *The False Presence of the Kingdom*, trans. C. E. Hopkin (New York: Seabury Press 1972), pp. 210f.: "I have few illusions.

In spite of all precautions, I know very well that [parts] will be used by devotees of the spiritual as a pretext for the cleavage between faith and life. I know very well that those same [parts] will be condemned by others as apolitical and pessimistic. I am fully aware that the [final] proposals . . . will be looked upon as ineffective and academic by those hungry for action, as superfluous by others, and as impractical by all."

(26) For a work which seeks to deal with the totality of Ellul's concerns, see Katharine C. Temple, *The Task of Jacques Ellul: A Proclamation of Biblical Faith as Requisite for Understanding the Modern Project* (unpublished Ph.D., McMaster University, 1976). I am indebted to Dr. Temple for much of what follows and wish to thank her both for permission to use her work and for the valuable insights about Ellul which I received from private conversations with her. I am especially grateful for some of the translations which she so kindly provided.

(27) Jacques Ellul, *The Technological Society,* trans. John Wilkinson (New York: Knopf, 1964). An updated version, *Le Système technicien* (Paris: Calmann-Lévy) was published in 1977 and translated into English in 1980: *The Technological System,* trans. J. Neugroschel (New York: Continuum).

(28) "Since 1935, I have been convinced that on the sociological plane, technique was by far the most important phenomenon, and that it was necessary to start from there to understand everything else." "From Jacques Ellul," *Katallagete* 2 (1970), p. 5.

(29) *Technological Society,* p. xxv.

(30) *Ibid.*, p. 38.

(31) *Ibid.*, p. 5.

(32) *Ibid.*, p. 130.

(33) See *ibid.*, p. 80: "[Man] is a device for recording effects and results obtained by various techniques. He does not make a choice of complex, and, in some way, human motives." Also, p. 135: "Man is reduced to the level of a catalyst. Better still, he resembles a slug inserted into a slot machine, he starts the operation without participating in it."

(34) *Ibid.*, p. 141.

(35) *Ibid.*, p. 58. See also *Metamorphose du Bourgeois* (Paris: Calmann-Levy, 1967), pp. 255f.

(36) *Metamorphose du Bourgeois*, pp. 218ff.

(37) See "Notes Préliminaires sur l'Eglise et Pouvoirs," *Foi et Vie* 73 (1974), p. 10.

(38) "La Technique et les Premiers Chapitres de la Genesis," *Foi et Vie* 59 (1960), pp. 112f.

(39) "A little Debate About Technology," *Christian Century* 90 (1973), p. 707.

(40) *The Meaning of the City,* trans. Dennis Pardee (Grand Rapids, Mich.: Wm. B. Eerdmanns, 1970), p. 21.

(41) *Hope in Time of Abandonment,* trans. C. E. Hopkin (New York: Seabury Press, 1973), p. 212.

(42) "La Technique," pp. 110f.

(43) *The Betrayal of the West,* trans. Matthew J. O'Connell (New York: Seabury, 1978), p. 137.

(44) Temple, *The Task of Jacques Ellul,* p. 440.

(45) See esp. *Betrayal of the West,* pp. 216f., and *Autopsy of Revolution,* trans. P. Wolf (New York: Knopf, 1971).

(46) "From Jacques Ellul," p. 5.

(47) " 'The World' in the Gospels," *Katallagete* 6 (1974), p. 19.

(48) See, e.g., "Notes Préliminaires," p. 4.

(49) George P. Grant, "A Platitude" in *Technology and Empire,* p. 137.

(50) The Josiah Wood Lectures (Sackville: Mount Allison University, 1974).

(51) *Ibid.*, p. 92. See also "Abortion and Rights" in *The Right to Birth: Some Christian Views on Abortion,* ed. Eugene Fairweather and Ian Gentiles (Toronto: the Anglican Book Centre, 1976), pp. 1-12. Written with Sheila Grant.

(52) Toronto: Copp Clark, 1959; 2nd ed. 1966.

(53) *Ibid.*, p. 29 (2nd ed.).

(54) See esp. *ibid.*, pp. 37ff.

(55) The book *George Grant in Process. Essays and Conversations* (Toronto: House of Anansi, 1978), ed. Larry Schmidt, has some interesting essays on the thought of Grant. Two in particular relate to our theme: William Mathie, "The Technological Regime: George Grant's Analysis of Modernity," pp. 157-166; and William Christian, "George Grant and the Terrifying Darkness," pp. 167-178.

(56) Grant is not the only one, of course, to lament the course of modern education. Perhaps one of the most interesting critiques comes from Arthur Koestler, who maintains that our educational institutions are dominated by the "three R's" of Reductionism, Ratomorphism, and Randomness. Koestler's concern with reductionism is well known and has been a major theme in his writings (cf. his novel

Arrival and Departure [London: Jonathan Cape, 1943], and also *Beyond Reductionism: The Alpbach Symposium 1968* [London: Hutchinson and Co., 1969], which he edited jointly with J. R. Smythies). Koestler finds it rather disconcerting to hear the human being described as "nothing but a complex biochemical mechanism, powered by a combustion system which energizes computers with prodigious storage facilities for retaining encoded information" ("Rebellion in a Vacuum," *Heel of Achilles: Essays 1968-1973* [London: Picador, 1976], p. 30.) Neither does he care for the description of the human being as a "naked ape" (as in Desmond Morris, *The Naked Ape: A Zoologist's Study of the Human Animal* [New York: McGraw-Hill, 1967]). Koestler maintains that reductionist thinking is very prevalent today.

Koestler coins the word "ratomorphism" to describe another prevalent view, which sees the behaviour of rats (or geese or pigeons) as a suitable model from which to extrapolate to the study of the human being. It is a view which is very popular today due to the immense influence of the discipline of psychology on our educational institutions, particularly at the higher levels. The influence of B. F. Skinner is all-pervasive here, of course, but others have also exercised influence -- for example, Konrad Lorenz and Desmond Morris. (It is worth noting here a fact which Koestler does not mention. Both Lorenz and Morris have described as "arrogant" the refusal of humans to believe that their knowledge of themselves is different from that of the animal kingdom -- see D. Morris, *The Naked Ape*, p. 14, and K. Lorenz, *On Aggression* [New York: Harcourt, Brace and World Inc., 1963], esp. pp. 192ff.)

"Randomness" is used by Koestler to describe the view that "biological evolution is considered to be nothing but random mutations preserved by natural selection; mental evolution nothing but random ties preserved by re-inforcement" (p.32). One only needs to observe some members of the modern school of art (those who beat material into random shapes) to appreciate how influential this idea has become. The despair evident in much modern literature and theatre is a natural concomitant of the random philosophy.

(57) "The University Curriculum" in *Technology and Empire*, pp. 113-133.

(58) *Ibid.*, p. 119.

(59) *Ibid.*, pp. 119f.

(60) *Ibid.*, p. 123.

(61) Historicism is a view which makes all comprehensive views relative. It is a view which is self-contradictory, for one cannot claim logically that all human thought is historically conditioned (and

therefore relative), without claiming universal (and therefore absolute) validity for the statement that all thought is relative. There is a more potent form of historicism, however, which meets the aforementioned objection by asseverating that there are a wide variety of comprehensive views, each as legitimate as the other. The horizon within which such views are formulated is imposed upon us by fate or circumstance. Radical historicism thus attempts to avoid the problem of asserting the universal validity of the insight that all thought is historical and relative. The historical character of thought may be maintained without having to transcend history. On this see the essay of Leo Strauss, "Natural Right and the Historical Approach" in *Political Philosophy*, esp. 148ff.

(62) Grant, "The University Curriculum," p. 131.

Notes to Conclusion to Part II

(1) Langdon Gilkey, *Understanding the Whirlwind,* pp. 40-61, also names these as the four key elements of the modern *Geist.*

(2) Cf. K. Jaspers, *Man in the Modern Age* (London: Routledge and Kegan Paul, 1933), p. 241: "The basic problem of our time is whether an independent, human being in his self-comprehended destiny is still possible. . . . Perhaps freedom has only existed for a real but passing moment between two immeasurably long periods of sleep, of which the first period was that of the life of nature and the second period was that of the life of technology. If so, human existence must die out . . . in a more radical sense than ever before."

Notes to Chapter Five

(1) For an excellent discussion of how the image "fellowship" functioned in the early church, see Ralph P. Martin, *The Family and the Fellowship: New Testament Images of the Church* (Grand Rapids, Mich.: Eerdmans, 1979).

(2) See above, p. 54.

(3) See Kueng, *On Being a Christian*, pp. 551ff.

(4) The debate evoked in England by the publication of *The Myth of God Incarnate*, ed. John Hick (London: S.C.M. Press, 1977), does not directly relate to the issues we are discussing. (In addition to *The Myth*, see its sequel *Incarnation and Myth*, ed. Michael Goulder [London: S.C.M. Press, 1979] and the replies in *The Truth of God Incarnate*, ed. Michael Green [London: Hodder and Stoughton, 1977].) While it is true that the problem of continuity and discontinuity is involved in the debate, the central issues of modernity, such as technique and human autonomy, are left out of account. This is perhaps because it is only in the North Amerian context -- where, as Grant has pointed out, modernity is manifested in a unique way -- that the real issues of modernity are fully perceived. The most interesting contributions to *The Myth* are by Dennis Nineham and Maurice Wiles. Both these men have had volumes of collected essays published in the *Explorations in Theology* series which are worth noting here: D. E. Nineham, *Explorations in Theology 1* (London: S.C.M. Press, 1977), esp. chs. 5, 6, 7, 8, 9, 10, and Maurice Wiles, *Explorations in Theology 4* (London: S.C.M. Press, 1979), esp. ch. 5. Among their many other works we should perhaps also single out M. Wiles, *The Remaking of Christian Doctrine* (London: S.C.M. Press, 1974); D. Nineham, *The Use and Abuse of the Bible: A Study of the Bible in an Age of Cultural Change* (London: Macmillan, 1976).

Another recent debate in England of some interest has centred around Edward Norman's B.B.C. Reith Lectures (1978), *Christianity and the World Order* (Oxford: Oxford University Press, 1979). This is a bold and thought-provoking book, which deals with the question of whether Christianity has become too politicized. Norman's work is, however, impaired by his setting up of too many straw men and his failure to give an adequate account of what he considers to be the essence of Christianity. There is an interesting (and hostile) reply to Norman by John J. Shepherd, "Norman Wisdom?" *ExTim* 90 (1979), 262-264.

(5) See above, ch. 2, n. 46.

(6) John Macquarrie, *Three Issues in Ethics* (New York: Harper and Row: 1970), p. 108.

(7) Thomas J. J. Altizer and William Hamilton, *Radical Theology and the Death of God*, (Indianapolis: Bobbs-Merrill, 1966), p. 15.

(8) *Pattern*, p. 36.

(9) H. E. W. Turner, "Orthodoxy and the Church Today," p. 167.

(10) It is interesting to note that there is a structural parallel to this in the argument among scientific circles about the nature of scientific progress. This debate was initiated by Thomas Kuhn in his *The Structure of Scientific Revolutions* (Chicago: The University of Chicago Press, 1962), where he argues that scientific progress is not evolutionary but revolutionary. Kuhn maintains that science progresses not by modifying and expanding what has gone before, but by replacing it with a completely different scientific world-view. Scientific paradigms embody not only scientific data but also methodological presuppositions. Thus, the paradigms of a particular scientific community determine the kinds of questions asked by that particular community and thus the method of inquiry and indirectly the results. Scientific communities, according to Kuhn, resist paradigm changes. Paradigm changes cannot be made one step at a time; they have to occur "all at once or not all" (p. 149). When they do occur, they are revolutionary because they totally transform ways of interpreting scientific data. Thus, in the paradigm shift from classical physics to relativity, the terms "time," "mass," and "velocity" acquired different meanings. There is a reply to Kuhn by Stephen E. Toulmin, "The Evolutionary Development of Natural Science," *American Scientist* 55 (1967), 456-471.

It is easy to see the structural similarity between Kuhn's argument and that of the theologians who urge that the religious sensibility of modern man is so different that terms such as "God," "sin," and "salvation," have acquired a completely different meaning. A criticism of Kuhn's work is that the criteria of judgment employed in adjudicating between paradigms are not (as Kuhn maintains) paradigm dependent. It is possible to judge between scientific paradigms by using common criteria: "There are criteria common to all paradigms of modern science -- criteria such as simplicity, coherence, and empirical agreement" (Ian G. Barbour, *Science and Secularity: The Ethics of Technology* [New York: Harper and Row, 1970], pp. 26f.). This is similar to the argument that we put forward in chapter two where it was argued that whether successive horizons within the

Christian tradition are intrinsically related must be judged according to criteria such as coherence and correspondence to the religious facts.

(11) *On Being a Christian*, p. 535.

(12) See "The Unconditioned in the Conditioned" in Kueng, pp. 536-539.

(13) See, e.g., *Philosophy in the Mass Age*, pp. 78ff.

(14) Kueng, *On Being a Christian*, p. 534: "Any assumption of meaning, truth, and rationality, of values and ideals, priorities and preferences, models and norms, presupposes a ***basic trust in reality.***"

(15) *Ibid.*, p. 536.

(16) Cf. Leon Kass, "The New Biology," n. 10: "The implicit goal of biomedical technology could well be said to be the reversal of the Fall and return of man to the hedonic and immortal existence of the Garden of Eden."

(17) See above, ch. 4, n. 15.

(18) "The Christian in a World of Technology," p. 262.

(19) See e.g., John Cogley, *Religion in a Secular Age* (London: Pall Mall, 1968), esp. Part II. Note also J. Metz, *Theology of the World* (New York: Herder & Herder, 1969).

(20) *Ibid.*, pp. 262f.

(21) *Ibid.*, p. 262.

(22) Evidence of technological feats in the ancient world abound; e.g., the excavations at Mohenjo-daro and Harappa. Katherine Kenyon's book *Digging Up Jericho* (London: Benn, 1957) shows the builders of this ancient city to have had a degree of sophistication in building which is quite remarkable. In the hellenistic world Archimedes, although he agreed with Plato that the philosopher should not put his knowledge to practical use, in fact made all kinds of mechanical devices -- see W. W. Tarn and G. T. Griffith, *Hellenistic Civilization* (London: Edward Arnold & Co., 1952), pp. 300f.

(23) *Ibid.*, pp. 263f.

(24) See J. Ellul, "La Technique et les premiers chapitres de la Genèse," *Foi et Vie* 59 (1960), 97-113. Ellul makes a special point of saying, however, that technology in itself is not evil -- see pp. 112f.

(25) "The Christian in a World of Technology," p. 264: "Another prerequisite for technological society is that man must believe change is both possible and desirable. He must have some reason for changing things, for initiating projects, for using tools to accomplish some purpose. The desire for change seems natural enough for us, but it would in no way appeal to a man trapped within a fatalistic or cyclical view of history. Only when history is seen as the theatre of

human response, as the scene of God's call and man's responsibility, does it make sense to try to alter what might have been. A 'closed universe' in which everything is already finished, simply waiting to be discovered by man, does not encourage technology."

(26) *Ibid.*

(27) Cox's actual words are: "Man was commissioned by God to name the animals and to tend the garden. Gerhard von Rad reminds us in his famous Genesis commentary that naming was for the Hebrews a kind of command, and that God himself had begun the creation by calling (that is, naming) the light 'day' and the darkness 'night'. Thus, man and God are both involved in the naming and controlling which constitute the creative process" (*ibid.*, p. 265). In fact, according to Genesis 1, creation and naming are not identical; cf. Genesis 1:3-5, where God creates light and darkness and *then* calls "the light Day, and the darkness he called Night." Cox goes on to argue that the rupture between God and humanity has been healed by the work of Christ, who has restored us to a position of responsibility and freedom, as Romans 5 and "the Galatians figure of sonship replacing tutelage to a schoolmaster" make clear. Cox's debt to Gogarten is explicit here.

(28) James Barr, "Man and Nature -- the Ecological Controversy and the Old Testament," *BJRL* 55 (1972-73), 9-32.

(29) Bertil Albrektson, *History and the Gods: An Essay on the Idea of Historical Events as Divine Manifestations in the Ancient Near East and in Israel,* Coniectanea Biblica Old Testament Series I (Lund: CWK Gleerup, 1967), p. 12. This is an important book which deserves to be read carefully by all who are interested in Old Testament theology.

(30) M. B. Foster, "The Christian Doctrine of Creation and the Rise of Modern Natural Science," *Mind* 43 (1934), 446-468; "Christian Theology and the Modern Science of Nature," *Mind* 44 (1935), 439-446, and 45 (1936), 1-27.

(31) Lynn White, "The Historical Roots of Our Ecologic Crisis," *Science* 155 (1967), 1203-1207.

(32) Claus Westermann, *The Genesis Accounts of Creation* (Philadelphia: Fortress, 1964), p. 22.

(33) Gerhard von Rad, *Genesis: A Commentary,* trans. John H. Marks (London: S.C.M. Press, 1961), p. 47.

(34) See the numerous examples of such an interpretation cited by J. Ellul, "La Technique," pp. 97ff.

(35) See, e.g., Jacob Jervell, *Imago Dei: Gen 1, 26. im Spaetjudentum, in der Gnosis und in der paulinischen Briefen,* Forschungen

zur Religion und Literatur des Alten und Neuen Testaments, NF 58 (Goettingen: Vandenhoeck and Ruprecht, 1960); Edvin Larsson, *Christus als Vorbild: Eine Untersuchung zu den paulinischen Tauf- und Eikontexten,* Acta Seminarii Neotestamentici Upsaliensis 23 (Uppsala: Almqvist and Wiksells, 1962); David Cairns, *The Image of God in Man* (London: S.C.M. Press, 1953).

(36) "God in Nature and in History," an ecumenical report published in *New Directions in Faith and Order: Bristol 1967* (Geneva: World Council of Churches, 1968), p. 18. Quoted by James Barr, "Man and Nature," p. 19, n. 3.

(37) "Man and Nature," p. 20.

(38) *Ibid.*, p. 19. See also J. Barr, "The Image of God in the Book of Genesis--A Study of Terminology," *BJRL* 51 (1968-69), 11-26.

(39) "Man and Nature," p. 20.

(40) *The Genesis Accounts,* p. 21.

(41) Von Rad, *Genesis,* p. 58.

(42) Although it is true that the story of the Fall (Genesis 3) does not belong to the P account, there is a recognition in P that the original harmony between God and humanity was disrupted by human sin, e.g., Gen 9:11f.

(43) *Genesis,* p. 74.

(44) Indeed, the classical political philosophers appear to have regarded the utilization of science to conquer nature as "unnatural" -- see George Grant, *Technology and Empire,* pp. 197ff.

(45) There is, of course, no semantic equivalent in Hebrew for the Greek *physis.* Instead we find an expression such as "all [God's] works." God is never differentiated from nature in the manner of the later Deists; neither is he identified with nature as in pantheism.

(46) This is the word used by C. F. D. Moule, *Man and Nature in the New Testament: Some Reflections on Biblical Ecology* (Philadelphia: Fortress, 1964), p. 3. Moule's book is interesting in that he thinks that the most satisfying interpretation of the image of God in humankind "is that which sees it basically as ***responsibility***" (p. 3).

(47) There is the possibility that *'elohim* are "angels." Cf. J. F. A. Sawyer, "The Meaning of *beselem 'elohim* ('in the image of God') in Genesis I-XI," *JTS* 25 (1975), pp. 423ff.

(48) *Earth Might Be Fair: Reflections on Ethics, Religion and Ecology,* ed. Ian G. Barbour (Englewood Cliffs: Prentice-Hall, 1972), Introduction, p. 7.

(49) John Macquarrie, "Creation and Environment," *ExTim* 83 (1971-72), 4-9, argues that for too long the dominant model for the

understanding of the relationship of God to the world has been a monarchical one: "God is a self-sufficient and transcendent being who creates the world by an act of will" (p. 6). Such a model is only a short step from Deism, in which the world is handed over to us to do with as we please. This, feels Macquarrie, is a drawback of modern secular theologies, and he appeals for a different understanding in which "God and the world are not sharply separated" (p. 6). This Macquarrie wishes to call the "organic model" and it is precisely what we are arguing for. See also Frederick Elder, *Crisis in Eden* (New York: Abingdon, 1970).

(50) Von Rad, *Genesis*, p. 75.

(51) See Perrin, *Kingdom*, pp. 69-73.

(52) This is not to say that Christianity stands entirely blameless for the ecological crisis. As Ian Barbour has said when speaking of the industrial development of the last few centuries: "The economic interests of the rising middle class, the competitiveness and rugged individualism of the capitalist ethos, the goals of economic productivity and efficiency -- aided, no doubt, by the 'Protestant ethic' of frugality, hard work and dominion over the earth -- all these encouraged a ruthlessness and arrogance towards nature unknown in earlier centuries" ("Attitudes toward Nature and Technology" in *Earth Might Be Fair*, p. 151). What we have argued is that there is no *integral* connection between the technological exploitation of nature and a Christian doctrine of creation. To derive technology from Christianity is to misunderstand the biblical view of creation, and to attribute the ecological crisis wholly to Christianity is to fail to do justice to a complex interplay of factors, to some of which Barbour alludes.

Notes to Conclusion

(1) S. Ogden, "Faith and Secularity," *God, Secularization and History: Essays in Honour of Ronald Gregor Smith,* ed. Eugene Long (Columbia: University of South Carolina Press, 1974), p. 30.

(2) Eugene Long, "God, Secularization and History," *ibid.*, p. 19.

(3) *Ibid.*, p. 14.

(4) *Three Issues in Ethics,* p. 108. Macquarrie formally defines this direction as "the direction which leads to a fuller humanity."

(5) Albert Schweitzer, *The Mystery of the Kingdom of God* (New York: Macmillan, 1950), p. 378.

(6) John V. Taylor, *Enough is Enough* (Minnesota: Augsburg, 1977), p. 56.

Bibliography

Albrektson, Bertil, *History and the Gods: An Essay on the Idea of Historical Events as Divine Manifestations in the Ancient Near East and in Israel*, Coniectanea Biblica Old Testament Series I, CWK Gleerup, Lund, 1967.

Allegro, John, *The Sacred Mushroom and the Cross*, Hodder & Stoughton, London, 1970.

Altizer, Thomas J. J., *The Gospel of Christian Atheism*, Westminster, Philadelphia, 1966.

------, *Mircea Eliade and the Dialectic of the Sacred*, Westminster, Philadelphia, 1963.

------, *Oriental Mysticism and Biblical Eschatology*, Westminster, Philadelphia, 1961.

------, *The Self-Embodiment of God*, Harper & Row, San Francisco, 1977.

Altizer, Thomas J. J., and William Hamilton, *Radical Theology and the Death of God*, Bobbs-Merrill, Indianapolis, 1966.

Bacon, F., *The Advancement of Learning*, ed. H. G. Dick, Random House, New York, 1955.

Barbour, Ian G. (ed.), *Earth Might Be Fair: Reflections on Ethics, Religion and Ecology*, Prentice-Hall, Englewood Cliffs, 1972.

------ (ed.), *Science and Religion: New Perspectives on the Dialogue*, Harper & Row, New York, 1968.

------, *Science and Secularity: The Ethics of Technology*, Harper & Row, New York, 1970.

Barr, James, "Man and Nature -- The Ecological Controversy and the Old Testament," *BJRL* 55 (1972-73), 9-32.

------, "The Image of God in the book of Genesis -- A Study of Terminology," *BJRL* 51 (1968-69), 11-26.

Barrett, C. K., "Albert Schweitzer and the New Testament," *ExpTim* 87 (1975), 4-10.

------, *Jesus and the Gospel Tradition*, S.P.C.K., London, 1967.

Barth, K., "Rudolf Bultmann -- An Attempt to Understand Him," in *Kerygma and Myth: A Theological Debate*, ed. H. W. Bartsch, S.P.C.K., London, 1962, pp. 83-132.

Bauer, Walter, *Orthodoxy and Heresy in Earliest Christianity*, ed. R. A. Kraft and G. Krodel, Fortress Press, Philadelphia, 1971.

Biemer, G., *Newman on Tradition*, Burns & Oates, London, 1967.

Black, J., *The Dominion of Man. The Search for Ecological Responsibility*, The University Press, Edinburgh, 1970.

Black, M., *An Aramaic Approach to the Gospels and Acts*, Clarendon Press, Oxford, 1953.

Bonhoeffer, Dietrich, *Letters and Papers From Prison*, S.C.M. Press, London, 1951.

Bornkamm, Guenther, *Jesus of Nazareth*, trans. Irene and Fraser McLuskey with James Robinson, Harper & Row, New York, 1960.

Braaten, Carl E., and Roy A. Harrisville, *The Historical Jesus and the Kerygmatic Christ: Essays on the New Quest of the Historical Jesus*, Abingdon, Nashville, 1964.

Brabazon, James, *Albert Schweitzer: A Biography*, G. P. Putnam's Sons, New York, 1975.

Brandon, S. G. F., *Jesus and the Zealots: A Study of the Political Factor in Primitive Christianity*, The University Press, Manchester, 1967.

Bultmann, Rudolph K., *The History of the Synoptic Tradition*, trans. John Marsh, Harper & Row, New York, 1968.

------, *Jesus and the Word*, trans. Louise Pettibone Smith and Erminie Huntress Lantero, Charles Scribner's Sons, New York, 1934, 1958.

------, *Jesus Christ and Mythology*, Charles Scribner's Sons, New York, 1958.

------, *Theology of the New Testament*, trans. K. Grobel, Charles Scribner's Sons, New York, 1951-55 (2 vols.).

Butterfield, Sir Herbert, *The Origins of Modern Science, 1300-1800*, rev. ed., Free Press, New York, 1957.

Cadbury, H. J., "Acts and Eschatology" in *The Background of the New Testament and its Eschatology. Studies in Honour of C. H. Dodd*, ed. W. D. Davies and D. Daube, Cambridge University Press, Cambridge, 1956, pp. 300-321.

Cairns, David, *The Image of God in Man*, S.C.M. Press, London, 1953.

Carmichael, Joel, *The Death of Jesus*, Macmillan, New York, 1962.

Carter, C. S., *A Hundred Years of Evolution*, Sidgwick & Jackson, London, 1957.

Cerfaux, L., *The Church in the Theology of St. Paul*, Herder, New York, 1963.

Chambers, Richard, *Vestiges of the Natural History of Creation*, J. Churchill, London, 1844.

Clarkson, Kathleen L., and David J. Hawkin, "Marx on Religion: The Influence of Bruno Bauer and Ludwig Feuerbach on His Thought and Its Implications for the Christian-Marxist Dialogue," *ScJTh* 31 (1978), 533-555.

Conzelmann, H., *An Outline of the Theology of the New Testament*, S.C.M. Press, London, 1969.

------, "Zur Analyse der Bekenntnisformel I Kor. 15,3-5," *Evangelische Theologie* 25 (1965), 1-11.

Cox, Harvey, "The Christian in a World of Technology," in *Science and Religion: New Perspectives on the Dialogue*, ed. Ian G. Barbour, Harper & Row, New York, 1968, pp. 261-280.

------, *God's Revolution and Man's Responsibility*, The Judson Press, Valley Forge, 1965.

------, *On Not Leaving It to the Snake*, S.C.M. Press, London, 1968.

------, *The Secular City: Secularization and Urbanization in Theological Perspective*, Macmillan, New York, 1965.

Cullmann, O., *The Early Church*, S.C.M. Press, London, 1956.

Dahl, N. A., "The Atonement -- An Adequate Reward for the Akedah? (Ro. 8:32)" in *Neotestamentica et Semitica. Studies in Honour of Matthew Black*, ed. E. Earle Ellis and Max Wilcox, T. & T. Clark, Edinburgh, 1969, pp. 15-29.

Darwin, Charles, *The Autobiography of Charles Darwin*, ed. Nora Barlow, Collins, London, 1958.

------, *The Descent of Man*, P. F. Collier & Son, New York, 1902.

------, *On the Origin of Species by Natural Selection, or the Preservation of Favoured Races in the Struggle for Life*, John Murray, London, 1859.

Davies, W. D., and David Daube (eds.), *The Background of the New Testament and its Eschatology*, Cambridge University Press, London, 1969.

Denney, J., *Jesus and the Gospel. Christianity Justified in the Mind of Christ*, Hodder & Stoughton, London, 1919.

Dibelius, M., *From Tradition to Gospel*, trans. Bertram Lee Woolf, Ivor Nicholson and Watson Ltd., London, 1934.

Dodd, C. H., *The Apostolic Preaching and its Developments*, Hodder & Stoughton, London, 1936.

------, "The Framework of the Gospel Narrative," *ExpTim* 43 (1932), 396-400.

------, *The Founder of Christianity*, Macmillan, New York, 1970.

Dungan, David L., "Mark -- The Abridgement of Matthew and Luke," in *Jesus and Man's Hope I,* Pittsburgh Theological Seminary, Pittsburgh, 1971, pp. 51-97.

Dunn, J. D. G., *Unity and Diversity in the New Testament,* S.C.M. Press, London, 1977.

Ehrhardt, A. A. T., "Christianity Before the Apostles' Creed," *HarvThR* 55 (1962), 73-119.

Eichrodt, Walter, *Theology of the Old Testament,* Vol. I, S.C.M. Press, London, 1961.

Eisler, Robert, *The Messiah Jesus and John the Baptist (According to Flavius Josephus' Recently Discovered "Capture of Jerusalem" and Other Jewish and Christian Sources*), Dial Press, New York, 1931.

Elder, F., *Crisis in Eden,* Abingdon, New York, 1970.

Ellul, Jacques, "A Little Debate About Technology," *Christian Century* 90 (1973), 706-707.

------, *Autopsy of Revolution,* trans. P. Wolf, Knopf, New York, 1971.

------, *The Betrayal of the West,* trans. Matthew J. O'Connell, Seabury, New York, 1978.

------, *The False Presence of the Kingdom,* trans. C. E. Hopkin, Seabury, New York, 1972.

------, "From Jacques Ellul," *Katallagete* 2 (1970), 5.

------, *Hope in Time of Abandonment,* trans. C. E. Hopkin, Seabury, New York, 1973.

------, "La Technique et les Premiers Chapitres de la Genesis," *Foi et Vie* 59 (1960), 97-113.

------, *The Meaning of the City,* trans. Dennis Pardee, Wm. B. Eerdmans, Grand Rapids, Mich., 1970.

------, *Métamorphose du Bourgeois,* Calmann-Lévy, Paris, 1967.

------, *The Technological Society,* trans. John Wilkinson, Knopf, New York, 1964. A much expanded and updated version has now been published under the title *The Technological System,* trans. J. Neugrosschel, Continuum, New York, 1980.

------, " 'The World' in the Gospels," *Katallagete* 6 (1974), 16-23.

Emmet, C. W., *The Eschatological Question in the Gospels, and Other Studies in Recent New Testament Criticism,* T. & T. Clark, Edinburgh, 1911.

Evans, C. F., "The Kerygma," *JTS* n.s. 7 (1965), 25-41.

Fine, Sidney, *Laissez Faire and the General Welfare State,* University of Michigan Press, Ann Arbour, 1964.

Foster, M. B., "The Christian Doctrine of Creation and the Rise of Modern Natural Science," *Mind* 43 (1934), 446-468.

------, "Christian Theology and the Modern Science of Nature," *Mind* 44 (1935), 439-466, and 45 (1936), 1-27.

Freud, Sigmund, *The Future of an Illusion*, trans. W. D. Robson-Scott, Doubleday, New York, 1961.

------, *The Interpretation of Dreams*, trans. & ed. James Strachey, Basic Books, New York, 1958.

------, *Moses and Monotheism*, trans. K. Jones, Vintage, New York, 1939.

------, *New Introductory Lectures on Psychoanalysis*, trans. W. J. H. Sprott, Hogarth Press, London, 1949.

------, "Obsessive Acts and Religious Practices" in *Sociology and Religion: A Book of Readings*, ed. Norman Birnbaum & G. Lenzer, Prentice-Hall, Englewood Cliffs, 1969, pp. 168-173.

------, *Totem and Taboo*, trans. A. A. Brill, Vintage Books, New York, 1918.

Funk, Robert W. (ed.), *Faith and Understanding*, Vol. I, trans. Louise Pettibone Smith, Harper & Row, New York, 1969.

Gerhardsson, B., *Memory and Manuscript: Oral Tradition and Written Transmission in Rabbinic Judaism and Early Christianity*, C. W. K. Gleerup, Lund, 1961.

------, *The Origins of the Gospel Traditions*, Fortress, Philadelphia, 1979.

Gilkey, Langdon, *Naming the Whirlwind: The Renewal of God-Language*, Bobbs-Merrill, Indianapolis, 1969.

Gogarten, F., *Despair and Hope For Our Time*, trans. Thomas Wieser, Pilgrim Press, Philadelphia, 1970.

------, *The Reality of Faith*, trans. Carl Michalson *et al.*, Westminster, Philadelphia, 1959.

Goulder, Michael (ed.), *Incarnation and Myth: The Debate Continued*, S.C.M. Press, London, 1979.

Graesser, Eric, *Das Problem der Parusieverzoegerung*, Toepelmann, Berlin, 1957.

Grant, George P., *English-Speaking Justice*, The Josiah Wood Lectures, Mount Allison University, Sackville, New Brunswick, 1974.

------, *Philosophy in the Mass Age*, Copp Clark, Toronto, 1959, 1966.

------, *Technology and Empire*, House of Anansi, Toronto, 1969.

------, *Time as History*, Canadian Broadcasting Corporation, Toronto, 1969.

------ and Sheila Grant, "Abortion and Rights" in *The Right to Birth: Some Christian Views on Abortion*, ed. Eugene Fairweather and Ian Gentiles, The Anglican Book Centre, Toronto, 1976, pp. 1-12.

Green, Michael (ed.), *The Truth of God Incarnate*, Hodder & Stoughton, London, 1977.

Hamilton, William, *The New Essence of Christianity*, Association Press, New York, 1961.

------, *God Is Dead*, Eerdmanns, Grand Rapids, Mich., 1966.

Harenberg, W. (ed.), *Der Spiegel on the New Testament*, Macmillan, New York, 1970.

Harnack, Adolf, *What Is Christianity?* Williams and Norgate, London, Edinburgh, and Oxford, 1901.

------, *Entstehung und Entwicklung der Kirchenfassung und des Kirchenrechts in den zwei ersten Jahrhunderten*, J. C. Hinrichs' Buchhandlung, Leipzig, 1910.

Hawkin, David J., "The Incomprehension of the Disciples in the Marcan Redaction," *JBL* 91 (1972), 491-500.

------"A Reflective Look at the Recent Debate on Orthodoxy and Heresy in Earliest Christianity," *Eglise et Théologie* 7 (1976), 367-378.

Hayes, John C., *Son of God to Superstar: Twentieth Century Interpretations of Jesus*, Abingdon, Nashville, 1976.

Heidegger, M. *Sein und Zeit*, Jahrbuch fuer Philosophie und phaenomenologische Forschung, Band viii, Max Niemeyer, Tuebingen, 1927.

------, *Basic Writings*, ed. D. F. Krell, Routledge & Kegan Paul, London, 1978.

Henderson, Ian, *Rudolf Bultmann*, Lutterworth Press, London, 1965.

Hengel, Martin, *Was Jesus a Revolutionist?* Fortress, Philadelphia, 1971.

Hick, John (ed.), *The Myth of God Incarnate*, S.C.M. Press, London, 1977.

Hiers, Richard H., *Jesus and Ethics: Four Interpretations*, Westminster, Philadelphia, 1968.

Hopper, David H., *A Dissent on Bonhoeffer*, Westminster, Philadelphia, 1975.

Horowitz, Gad, "Red Tory" in *Canada: A Guide to the Peaceable Kingdom*, ed. William Kilbourn, Macmillan, Toronto, 1970, pp. 254-260.

Hunter, A. M., *The Unity of the New Testament*, S.C.M. Press, London, 1943.

Jaspers, Karl, *Man in the Modern Age*, Routledge & Kegan Paul, London, 1933.

------ and R. Bultmann, *Myth and Christianity: An Inquiry Into the Possibility of Religion Without Myth*, Noonday Press, New York, 1958.

Jeremias, J., "Artikelloses Christos. Zur Ursprache von I Kor. XV 3b-5," *ZNW* 57 (1966), 211-215.

Jervell, Jacob, *Imago Dei: Gen 1, 26f. im Spaetjudentum, in der Gnosis und in den paulinischen Briefen*, Forschungen zur Religion und Literatur des Alten und Neuen Testaments, Neue Folge 58, Vandenhoeck & Ruprecht, Goettingen, 1960.

Jones, G. V., *Christology and Myth in the New Testament*, Allen & Unwin, London, 1956.

Kaesemann, E., "The Canon of the New Testament and the Unity of the Church" in his *Essays on New Testament Themes*, S.C.M. Press, London, 1965, pp. 95-107.

------, "The Problem of the Historical Jesus" in *Essays on New Testament Themes*, pp. 15-47.

------, *The Testament of Jesus: A Study of the Gospel of John in the Light of Chapter 17*, S.C.M. Press, London, 1968.

Kass, Leon R., "The New Biology: What Price the Relieving of Man's Estate?" in *Science, Technology and Freedom*, ed. Willis H. Truitt and T. W. Graham Solomons, Houghton Mifflin Co., Boston, 1974, pp. 149-169.

Kaufman, Walter, *The Portable Nietzsche*, Viking Press, New York, 1968.

Kegley, Charles W. (ed.), *The Theology of Rudolf Bultmann*, Harper & Row, New York, 1966.

Kelly, J. N. D., *Early Christian Creeds*, Longmans, London, 1960.

------, *Early Christian Doctrines*, 2nd ed., Adam & Charles Black, London, 1960.

Kenyon, Katherine, *Digging up Jericho*, Benn, London, 1957.

Klappert, B., "Zur Frage des semitischen der greichischen Urtextes von I Kor. XV 3-5," *NTS* 13 (1967), 168-173.

Kluckholn, Clyde, "Recurrent Themes in Mythology" in *The Making of Myth*, ed. Richard Malin Ohmann, Putnam, New York, 1962, pp. 53-63.

Knox, Sir Malcolm, *A Layman's Quest*, George Allen & Unwin, London, 1969.

Knox, W. L., *The Acts of the Apostles,* Cambridge University Press, Cambridge, 1948.

Koester, H., "GNOMAI DIAPHORAI: The Origin and Nature of Diversification in Early Christianity," *HarvThR* 58 (1965), 279-318.

------, "The Theological Aspects of Primitive Christian Heresy" in *The Future of Our Religious Past,* ed. James M. Robinson, Harper & Row, New York, 1971, pp. 65-83.

Koestler, Arthur, *The Ghost in the Machine,* Hutchinson, London, 1967.

------, *The Heel of Achilles: Essays 1968-1973,* Picador, London, 1976.

------ and J. R. Smythies (ed.), *Beyond Reductionism: The Alpbach Symposium 1968,* Hutchinson, London, 1969.

Kroeber, A. C. "Totem and Taboo: An Ethnologic Analysis" in *A Reader in Comparative Religion,* ed. W. A. Lessa & E. Z. Vogt, Harper & Row, New York, 1965, 45-53.

Kuemmel, W. G., *The New Testament: The History of the Investigation of its Problems,* S.C.M. Press, London, 1973.

Kueng, Hans, *On Being a Christian,* trans. Edward Quinn, Collins, London, 1974.

Kuhn, Thomas, *The Structure of Scientific Revolutions,* University of Chicago Press, Chicago, 1962.

Larsson, Edvin, *Christus als Vorbild: Eine Untersuchung zu den paulinischen Tauf- und Eikontexten,* Acta Seminarii Neotestamentici Upsaliensis 23, Almqvist & Wiksells, Uppsala, 1962.

Lewis, C. S., *They Asked For a Paper,* Geoffrey Bles, London, 1962.

Linton, O. *Das Problem der Urkirche in der neueren Forschung: eine kritische Darstellung,* Almqvist & Wiksells, Uppsala, 1932.

Lightfoot, R. H., *History and Interpretation in the Gospels,* Hodder & Stoughton, London, 1935.

------, *The Gospel Message of Mark,* Clarendon, Oxford, 1950.

Livingston, James C., *Modern Christian Thought. From the Enlightenment to Vatican II,* Macmillan, New York, 1971.

Lloyd, G. E. R., *Early Greek Science: Thales to Aristotle,* Chatto & Windus, London, 1970.

Lonergan, Bernard J. F., *Method in Theology,* Darton, Longman & Todd, London, 1972.

Long, Eugene, *God, Secularization and History: Essays in Memory of Ronald Gregor Smith,* University of South Carolina Press, Columbia, 1974.

Lorenz, Konrad, *On Aggression,* Harcourt, Brace & World, New York, 1963.

Machiavelli, Niccoló, *The Discourses,* edited with introduction by Bernard Crick, using the translation of Leslie Walker, with revisions by Brian Richardson, Penguin, London, 1970.

------, *The Prince,* trans. Robert M. Adams, W. W. Norton & Co. Inc., New York, 1977.

MacIntryre, Alasdair, *Difficulties in Christian Belief,* S.C.M. Press, London, 1959.

Macquarrie, John, *Three Issues in Ethics,* Harper & Row, New York, 1970.

------, *Paths in Spirituality,* Harper & Row, New York, 1972.

------, *The Scope of Demythologizing: Bultmann and His Critics,* Harper & Row, New York, 1961.

------, "Creation and Environment," *ExpTim* 83 (1971-72), 4-9.

Manson, T. W., *The Teaching of Jesus: Studies of its Form and Content,* 2nd ed., Cambridge University Press, Cambridge, 1935.

Marx, Karl, *Capital,* ed. F. Engels, trans. Samuel Moore and Edwood Avelling, The Modern Library, New York, 1906.

------, "Contribution to the Critique of Hegel's *Philosophy of Right*: Introduction" in *Early Writings,* trans. T. B. Bottomore, McGraw-Hill, New York, 1964, pp. 41-59.

------, *The Difference Between the Democritean and Epicurean Philosophy of Nature,* trans. N. Livergood, Martinus Nijhoff, The Hague, 1967.

------, "Excerpt-Notes of 1884" in *Writings of the Young Marx on Philosophy and Society,* trans. Lloyd Easton and Kurt Guddat, Doubleday, New York, 1967, pp. 265-282.

------, *The Holy Family,* trans. R. Dixon, Foreign Languages Publishing House, Moscow, 1956.

------ and F. Engels, *The German Ideology,* trans. S. Ryazanskaya, Progress Publishers, Moscow, 1964.

Martin, Ralph, *The Family and the Fellowship: New Testament Images of the Church,* Eerdmans, Grand Rapids, Mich., 1979.

Mayor, S. H., "Jesus and the Christian Understanding of Society," *ScJTh* 32 (1979), 45-60.

McKenzie, A. E. E., *The Major Achievements of Science,* Vol. I, Cambridge University Press, Cambridge, 1967.

Metz, Johannes, *Theology of the World,* Herder & Herder, New York, 1969.

Meyer, B. F., *The Church in Three Tenses*, Doubleday, Garden City, New York, 1971.

------, *The Aims of Jesus*, S.C.M. Press, London, 1979.

Miller, P., and K. Pound, *Creeds and Controversies*, English Universities Press Ltd., London, 1969.

Miller, W. R., *The New Christianity*, Dell, New York, 1967.

Moltmann, J., *Religion, Revolution and the Future*, Scribner's, New York, 1969.

Morris, Desmond, *The Naked Ape: A Zoologist's Study of the Human Animal*, McGraw-Hill, New York, 1967.

Moule, C. F. D., *Man and Nature in the New Testament: Some Reflections on Biblical Ecology*, Fortress, Philadelphia, 1964.

Neill, Stephen, *The Interpretation of the New Testament 1861-1961*, Oxford University Press, London, 1964.

Newman, J. H., *An Essay on the Development of Christian Doctrine*, Sheed & Ward, London, 1960.

------, *Oxford University Sermons*, Rivingtons, London, 1892.

Niebuhr, Reinhold (ed.), *Marx and Engels on Religion*, Schocken Books, New York, 1964.

Nietzsche, F., *The Use and Abuse of History*, trans. Adrian Collins, Bobbs-Merrill, Indianapolis & New York, 1957.

Nineham, D. E., "Schweitzer Revisited" in *Explorations in Theology I*, S.C.M. Press, London, 1977, pp. 112-133.

------, "The Order of Events in St. Mark's Gospel -- An Examination of Dr. Dodd's Hypothesis" in *Studies in the Gospels*, ed. D. E. Nineham, Blackwell, Oxford, 1953, pp. 223-240. Reprinted in *Explorations in Theology I*, pp. 7-23.

------, *The Use and Abuse of the Bible: A Study of the Bible in an Age of Cultural Change*, Macmillan, London, 1976.

Nordern, E., *Agnostos Theos. Untersuchung zur Formgeschichte religioser Rede*, Teubner, Stuttgart, 1913.

Norman, Edward, *Christianity and the World Order*, Oxford University Press, Oxford, 1979.

Perrin, N., *Jesus and the Language of the Kingdom,*, Fortress, Philadelphia, 1976.

------, *The Kingdom of God in the Teaching of Jesus*, S.C.M. Press, London, 1963.

------, *What Is Redaction Criticism?* Fortress, Philadelphia, 1969.

Ray, R. R.., "Jacques Ellul's Innocent Notes on Hermeneutics," *Interpretation* 33 (1979), 268-282.

Reimarus, H. S., *On the Intention of Jesus and His Disciples* in *Reimarus: Fragments*, ed. Charles Talbert, trans. Ralph S. Fraser, Fortress, Philadelphia, 1970.

Rieff, Philip, *Freud: The Mind of the Moralist*, Doubleday, New York, 1959.

Riesenfeld, Harald, *The Gospel Tradition and its Beginnings. A Study in the Limits of "Formgeschichte,"* Mowbray, London, 1957.

Robinson, James M., "Basic Shifts in German Theology," *Interpretation* 16 (1962), 76-97.

------, Review of R. Bultmann, *Theology of the New Testament II*, *Theology Today* 13 (1956-57), 261-269.

------, *The Problem of History in Mark*, Studies in Biblical Theology 21, S.C.M. Press, London, 1957.

------, *A New Quest of the Historical Jesus*, S.C.M. Press, London, 1959.

Robinson, J. A. T., *Honest to God*, S.C.M. Press, London, 1959.

Robertson, John C., "Hermeneutics of Suspicion *versus* Hermeneutics of Good Will," *Studies in Religion* 8 (1979), 365-377.

Rogerson, J. W., *Myth in Old Testament Interpretation*, BZAW 134, De Gruyter, Berlin & New York, 1974.

------, "Slippery Words: V. Myth," *ExpTim* 90 (1978), 10-14.

Rubenstein, Richard L., *After Auschwitz: Radical Theology and Contemporary Judaism*, Bobbs-Merrill, Indianapolis, 1966.

Runner, H. Evan, *Scriptural Religion and Political Task*, Wedge Publishing Foundation, Toronto, 1974.

Sanders, E. P., *The Tendencies of the Synoptic Tradition*, Cambridge University Press, Cambridge, 1969.

Sanders, Jack T., *Ethics in the New Testament: Change and Development*, Fortress, Philadelphia, 1975.

Sawyer, J. F. A., "The Meaning of *beselem 'elohim* ('in the image of God') in Genesis I-XI," *JTS* 25 (1975), 418-426.

Schilling, Paul S., *Contemporary Continental Theologians*, S.C.M. Press, London, 1966.

Schmidt, K. L., *Die Rahmen der Geschichte Jesu*, Trowitsch & Sohn, Berlin, 1919.

Schmidt, Larry, *George Grant in Process. Essays and Conversations*, House of Anansi, Toronto, 1978.

Schweitzer, Albert, *The Mystery of the Kingdom of God,* trans. Walter Lowrie, A. & C. Black, London, 1914; Macmillan, London, 1950.

------, *Out of My Life and Thought: An Autobiography,* trans. C. T. Campion, Holt, Rinehart & Winston, New York, 1933.

------, *The Psychiatric Study of Jesus,* trans. Charles R. Joy, Beacon Press, Boston, 1948.

------, *The Quest of the Historical Jesus: A Critical Study of its Progress From Reimarus to Wrede,* trans. W. Montgomery, A. & C. Black, London, 1910; 1936.

Shiner, Larry, *The Secularization of History: An Introduction to the Theology of Friedrich Gogarten,* Abingdon, New York, 1966.

Skinner, Quentin, *Machiavelli,* Cambridge University Press, Cambridge, 1981.

Snell, Bruno, *The Discovery of the Mind: The Greek Origins of European Thought,* Harvard University Press, Cambridge, Mass., 1953.

Sohm, R., *Kirchenrecht,* Vol. I, Duncker & Humblet, Leipzig, 1892.

Stanley, D. M., "The Conception of Salvation in Primitive Christian Preaching," *CBQ* 18 (1956), 231-254.

Strauss, David Friedrich, *The Life of Jesus Critically Examined,* trans. from the 4th ed. (1840) by Marion Evans, Calvin Blanchard, New York, 1860; republished in 1970 by Scholarly Press, Michigan.

Strauss, Leo, *Natural Right and History,* University of Chicago Press, Chicago, 1953.

------, *Thoughts on Machiavelli,* University of Washington Press, Seattle & London, 1958.

------, "The Three Waves of Modernity" in *Political Philosophy; Six Essays by Leo Strauss,* ed. Hilail Gildin, Pegasus, New York & Indianapolis, 1975, pp. 81-98.

Sykes, S. W., and J. P. Clayton (ed.), *Christ, Faith and History,* Cambridge University Press, Cambridge, 1972.

Talbert, Charles H. (ed.), *Reimarus: Fragments,* trans. Ralph S. Fraser, Fortress, Philadelphia, 1970.

Tarn, W. W., and G. T. Griffith, *Hellenistic Civilization,* 3rd ed., Edward Arnold & Co., London, 1952.

Taylor, John V., *Enough is Enough,* Augsburg, Minnesota, 1977.

Taylor, V., *The Formation of the Gospel Tradition,* Macmillan, London, 1935.

Tawney, R. H., *Religion and the Rise of Capitalism: An Historical Study,* Mentor, New York & Toronto, 1954.

Temple, Katherine C., "The Task of Jacques Ellul: A Proclamation of Biblical Faith as Requisite for Understanding the Modern Project," unpublished Ph.D. dissertation, McMaster University, 1976.

Toulmin, Stephen E., "The Evolutionary Development of Natural Science," *American Scientist* 55 (1967), 456-457.

Trocmé, Etienne, *Jesus and His Contemporaries,* S.C.M. Press, London, 1973.

Turner, H. E. W., *The Pattern of Christian Truth: A Study in the Relations Between Orthodoxy and Heresy in the Early Church,* Bampton Lectures, Mowbray, London, 1954.

------, "Orthodoxy and the Church Today," *The Churchman* 86 (1972), 166-173.

van Buren, Paul, *Discerning the Way: A Theology of the Jewish-Christian Reality,* Seabury, New York, 1980.

------, *The Secular Meaning of the Gospel,* Macmillan, New York, 1963.

van Leeuwen, Arend, *Christianity in World History,* Scribner's, New York, 1965.

von Rad, Gerhard, *Genesis: A Commentary,* trans. H. Marks. S.C.M. Press, London, 1961.

Weber, Max, *The Protestant Ethic and the Spirit of Capitalism,* trans. Talcott Parsons, Scribner's, New York, 1958.

Weiss, Johannes, *Jesus' Proclamation of the Kingdom of God,* trans. and ed. Richard H. Hiers and David L. Holland, Fortress, Philadephia, 1971.

Westermann, Claus, *The Genesis Accounts of Creation,* Fortress, Philadelphia, 1964.

White, Lynn, Jr., "The Historical Roots of Our Ecologic Crisis," *Science* 155 (1967), 1203-1207.

Whitfield, J. H., *Discourses on Machiavelli,* Heffer, Cambridge, 1969.

Wilberforce, Samuel, "Darwin's *Origin of Species,*" *Quarterly Review* 108 (1860), 225-264.

Wilcox, Max, *The Semitisms of Acts,* Clarendon, Oxford, 1965.

Wiles, M. F., "Does Christology Rest on a Mistake?" *Religious Studies* 6 (1970), 69-76.

------, "In Defence of Arius," *JTS* 13 (1962), 339-347.

------, *Explorations in Theology 4,* S.C.M. Press, London, 1979.

------, *The Remaking of Christian Doctrine*, The Huslean Lectures 1973, S.C.M. Press, London, 1974.

Wrede, William, *The Messianic Secret*, trans. J. C. G. Greig, J. Clark, Cambridge, 1971.

Yoder, John Howard, *The Politics of Jesus*, William B. Eerdmans, Grand Rapids, Mich., 1972.

Index

Adam, 88, 105, 107, 111f.
Alienation, 71, 89, 104
Apocalyptic, 16, 18, 19, 22, 32, 45, 118
Aquinas, 60f.
Aristotle, 60f., 73, 80, 145n. 5
Augustine, 89, 115
Autonomy, ix, 80-93, 94f., 103-106, 114, 116f.

Bacon, Francis, 60f., 74f., 145n. 7, 148n. 52, 150n. 2
Barr, James, 108, 112, 160n. 28, 161n. 38
Barrett, C. K., vii, 55, 127n. 66, 144n. 13
Barth, K., 29, 129n. 104
Bauer, Walter, 35-40, 100f., 132n. 5, 133n. 6, 134nn. 24,32
Berkhof, H., 111
Biblical doctrine of creation, 106-109, 110
Biblical revelation, 36, 101
Bonhoeffer, D., 84f., 151nn. 18,20
Bornkamm, G., 30, 126n. 39, 130n. 110
Bultmann, R., 3, 11, 15, 20-30, 31ff., 49, 53, 81, 101, 127nn. 74,77,78, 128nn. 80,84,85,86, 89,93,94, 129nn. 97,98,102, 130nn. 112,113, 133n. 6

Calvin, John, 80
Canon, 10, 137n. 47
Catastrophism, 62
Catholic Worker, 119
Chambers, Richard, 62
Clarkson, Kathleen, vii, 147n. 38
Classical political philosophy, 73f.
Continuity and discontinuity, 99-103
Copernicus, 60, 64
Cox, Harvey, 83, 105f., 109, 150n. 13, 151n. 15,16, 160n. 27
Creeds, 36, 43, 47
Cuvier, Baron, 62

Darwin, Charles, 61-65, 72, 73, 79, 80, 94, 145nn. 8,9
Deism, 71
Descartes, R., 81
Development, 34ff., 39, 42, 43-48, 49f., 82f., 83, 86, 100-102, 137n. 49, 138n. 57
Dodd, C. H., 43, 47, 125n. 45, 139n. 63
Dibelius, M., 21, 23, 127n. 73

Eichrodt W., 50, 141n. 68
Ellul, Jacques, 85-90, 93f., 102f., 105, 109, 116, 120, 149n. 64, 151nn. 24,25,26,27,28,33, 159n. 24
Eschatology, 14, 16, 26, 31, 36, 125
Eusebian view of history, 34
Eve, 105
Evolution, 61, 63f., 69
Existentialism, 25f., 29, 33, 36, 45, 101, 115, 118

Faith formulas, 47ff.
Fall, the, 64, 88f., 105, 107, 113f., 159n. 16, 161n. 42
False consciousness, 71
Fixed and flexible elements (in early Christianity), 36, 40, 43ff., 54, 100ff., 117
Form criticism, 11, 18, 22ff.

Foster, M. B., 108, 160n. 30
Freud, S., 21, 61, 65-69, 72, 73, 79, 80, 94, 146n. 22, 147nn. 24,29,32

Galileo, 61
Garden of Eden, 89, 105, 107f., 112f., 159n. 16
Genesis, 63, 64, 81, 88, 105-111, 113ff., 160n. 27
Gogarten, F., 80ff., 103, 117, 150nn. 3,4,6,9,12,14, 160n. 27
Grant, G. P., 78, 90-93, 94, 101ff., 105, 118, 120, 149nn. 61,66, 153nn. 49,51, 55,56, 161n. 44

Harnack, A., 12, 20, 21, 35
Hegel, F., 8, 10, 70, 71, 147n. 38
Heidegger, M., ix, 25, 29, 128n. 86
Heresy, 34-42, 102, 132nn. 4,5, 134nn. 24,32, 137n. 47
Heteronomous ethic, 54
Heteronomous theology, 119
Historicism, 92, 94, 154n. 61
Holy Spirit, 41, 42, 55, 102
Huxley, Aldous, 78, 105
Huxley, Thomas, 64

Ideology, 71f., 88

John, Gospel of, 6, 14, 28, 140n. 63

Kaesemann, E., 29f., 129n. 107, 132n. 5, 133n. 6
Kass, Leon, 79, 149n. 58, 159n. 16
Kelly, J. N. D., 48, 139n. 62
Kerygma, 25, 27, 30, 36, 43, 49, 139n. 63
Kingdom of God, 5, 6, 12, 14, 16, 26, 31, 125n. 39, 130n. 111
Koester, H., 38, 40, 41, 43, 46, 133n. 6, 135n. 35, 137n. 49
Koestler, A., 59f., 145n. 2, 153n. 56
Kuemmel, G., 4, 123n. 7,8
Kueng, H., 20, 54, 104, 127n. 68, 157n. 3, 159n. 14
Kuhn, T., 158n. 10

Last men, 76f.
Latent content (of dreams), 66, 68
Lewis, C. S., 59, 145n. 1
Lex credendi, 41
Lex orandi, 36, 40-51, 99ff., 137n. 47
Liberalism, 78f., 92, 109
Lonergan, B., 101, 129n. 101, 134nn. 29,30,31, 135n. 36
Long, E., 118
Lorenz, Konrad, 153n. 56
Luther, M., 55, 60, 80, 82
Lyell, Charles, 62

MacIntyre, A., 69, 147n. 33
Machiavelli, N., 73-75, 77, 79, 80, 94, 104, 148nn. 49,50,51, 53,55, 150n. 2
Macquarrie, J., 50, 101, 105, 118, 129n. 105, 141n. 69, 158n. 6, 161n. 49, 163n. 4
Madman, parable of, 76
Malthus, T., 63
Manifest content (of dreams), 66
Manson, T. W., 22, 127n. 75
Marcion, 35, 39
Marx, K., 61, 69-72, 73, 79, 80, 94, 147nn. 38,39, 140nn. 40,45
Messiah, 6, 7, 15, 17, 45
Meyer, B. F., vii, 44, 137n. 48, 138n. 57
Morris, Desmond, 153n. 56
Myth, 9-12, 26ff., 83, 124n. 23, 129n. 97, 157n. 4

Natural law, 80, 90f., 99, 101, 103, 105
Natural selection, 62f.
Newman, J. H., 136n. 45
Newton, Isaac, 61, 64
Nicaea, 42, 46
Nietzsche, F., 75-77, 78f., 80, 84, 91, 94, 104, 149n. 59
Nihilists, 76, 77
Noah, 111-113
Norman, Edward, 157n. 4
Nuclear weapons, 119

Oedipus Complex, 67ff.
Ontology, 36, 118
Origen, 41, 137n. 47
Orthodoxy, 34-51, 104, 132n. 45, 134nn. 24,32, 137n. 47

Paul, 27, 28, 34, 44, 46, 47, 48, 82, 103, 116, 120, 139nn. 61,63
Perrin, N., 12, 19, 125nn. 33,42,43, 130n. 111, 143n. 12, 162n. 51
Pietism, 99
Protestantism, 71, 80

Reductionism, 79, 154
Reimarus, H. S., 3, 4-7, 11, 13, 20, 31, 32, 123n. 14
Relativism, 79, 94, 118
Ritschl, A., 12, 14
Robinson, J. A. T., 54, 143n. 9
Rule of faith, 36, 43, 48, 101

Schilling, Paul, 82, 83, 150nn. 8,12
Schweitzer, Albert, 3, 4, 7, 8, 11, 12, 13-20, 31, 32, 52f., 101, 120, 123n. 3, 125nn. 35,36, 37,38, 163n. 5
Science, 9, 26, 60f., 75, 79, 80, 84, 85, 92, 94, 106ff., 120, 145n. 3, 158n. 10, 160n. 30, 161n. 44
Scientific Revolution, 60f., 61, 94
Secularism, 81f., 95, 150n. 14
Secularization, 61, 81f., 95, 150nn. 13,14, 163n. 1
Shiner, L., 81, 150nn. 4,5,9
Socrates, 39, 76
Son of God, 6, 45
Son of Man, 16, 17, 19, 126n. 52
Strauss, D. F., 3, 4, 7, 8-12, 20, 31, 32, 124nn. 21,30, 131n. 115
Strauss, Leo, 148nn. 47,51,53,57, 150n. 2, 154n. 61
Strecker, G., 38, 132n. 4
Supranaturalism, 11
Supranaturalist ethic, 54
Syncretism, 43f., 100

Taylor, J. V., 120, 163n. 6
Technique, 79, 85-93, 152n. 28
Technology, 78, 80-93, 94f., 106-116, 118ff., 153nn. 39,55, 159nn. 22,24,25, 162n. 52
Teleology, 62
Temple, Katharine, vii, 152nn. 26,44
Toulmin, S., 158n. 10
Turner, H. E. W., 35-38, 40, 41, 43-46, 49, 100-102, 133n. 21, 136n. 44, 137nn. 47,49, 143n. 8, 158n. 9

Unconscious, theory of, 65f.
University curriculum, 92, 154n. 57

Von Rad, G., 112f., 160n. 33, 162n. 50

Wallace, A., 63, 146n. 16
Weber, Max, 80
Weiss, J., 14, 15, 16, 21, 26, 31, 125n. 39

Westermann, Claus, 110, 112, 160n. 32
White, Lynn, 109, 115, 160n. 31
Whitfield, J. H., 74, 148n. 50
Wilberforce, S., 64, 146n. 17
Winiarski, W., 75, 148n. 56
Wrede, W., 13ff., 21, 125nn. 40,45

SR SUPPLEMENTS

1. **FOOTNOTES TO A THEOLOGY**
   The Karl Barth Colloquium of 1972
   Edited and Introduced by Martin Rumscheidt
   1974 / viii + 151 pp.
2. **MARTIN HEIDEGGER'S PHILOSOPHY OF RELIGION**
   John R. Williams
   1977 / x + 190 pp.
3. **MYSTICS AND SCHOLARS**
   The Calgary Conference on Mysticism 1976
   Edited by Harold Coward and Terence Penelhum
   1977 / viii + 121 pp. / OUT OF PRINT
4. **GOD'S INTENTION FOR MAN**
   Essays in Christian Anthropology
   William O. Fennell
   1977 / xii + 56 pp.
5. **"LANGUAGE" IN INDIAN PHILOSOPHY AND RELIGION**
   Edited and Introduced by Harold G. Coward
   1978 / x + 98 pp.
6. **BEYOND MYSTICISM**
   James R. Horne
   1978 / vi + 158 pp.
7. **THE RELIGIOUS DIMENSION OF SOCRATES' THOUGHT**
   James Beckman
   1979 / xii + 276 pp. / OUT OF PRINT
8. **NATIVE RELIGIOUS TRADITIONS**
   Edited by Earle H. Waugh and K. Dad Prithipaul
   1979 / xii + 244 pp. / OUT OF PRINT
9. **DEVELOPMENTS IN BUDDHIST THOUGHT**
   Canadian Contributions to Buddhist Studies
   Edited by Roy C. Amore
   1979 / iv + 196 pp.
10. **THE BODHISATTVA DOCTRINE IN BUDDHISM**
    Edited and Introduced by Leslie S. Kawamura
    1981 / xxii + 274 pp.
11. **POLITICAL THEOLOGY IN THE CANADIAN CONTEXT**
    Edited by Benjamin G. Smillie
    1982 / xii + 260 pp.
12. **TRUTH AND COMPASSION**
    Essays on Judaism and Religion in Memory of Rabbi Dr. Solomon Frank
    Edited by Howard Joseph, Jack N. Lightstone, and Michael D. Oppenheim
    1983 / vi + 217 pp.
13. **CRAVING AND SALVATION**
    A Study in Buddhist Soteriology
    Bruce Matthews
    1983 / xiv + 138 pp.
14. **THE MORAL MYSTIC**
    James R. Horne
    1983 / x + 134 pp.
15. **IGNATIAN SPIRITUALITY IN A SECULAR AGE**
    Edited by George P. Schner
    1984 / viii + 128 pp.
16. **STUDIES IN THE BOOK OF JOB**
    Edited by Walter E. Aufrecht
    1985 / xii + 76 pp.
17. **CHRIST AND MODERNITY**
    Christian Self-Understanding in a Technological Age
    David J. Hawkin
    1985 / x + 188 pp.

EDITIONS SR

1. **LA LANGUE DE YA'UDI**
   Description et classement de l'ancien parler de Zencirci dans le
   cadre des langues sémitiques du nord-ouest
   Paul-Eugène Dion, O.P.
   1974 / viii + 511 p.
2. **THE CONCEPTION OF PUNISHMENT IN EARLY INDIAN LITERATURE**
   Terence P. Day
   1982 / iv + 328 pp.

3. **TRADITIONS IN CONTACT AND CHANGE**
   Selected Proceedings of the XIVth Congress of the International Association for the History of Religions
   Edited by Peter Slater and Donald Wiebe with Maurice Boutin and Harold Coward
   1983 / x + 758 pp.
4. **LE MESSIANISME DE LOUIS RIEL**
   Gilles Martel
   1984 / xviii + 484 p.
5. **MYTHOLOGIES AND PHILOSOPHIES OF SALVATION IN THE THEISTIC TRADITIONS OF INDIA**
   Klaus K. Klostermaier
   1984 / xvi + 552 pp.
6. **AVERROES' DOCTRINE OF IMMORTALITY**
   A Matter of Controversy
   Ovey N. Mohammed
   1984 / vi + 202 pp.
7. **L'ETUDE DES RELIGIONS DANS LES ECOLES**
   L'expérience américaine, anglaise et canadienne
   Fernand Ouellet
   1985 / xvi + 672 p.

STUDIES IN CHRISTIANITY AND JUDAISM / ETUDES SUR LE CHRISTIANISME ET LE JUDAISME

1. **A STUDY IN ANTI-GNOSTIC POLEMICS**
   Irenaeus, Hippolytus, and Epiphanius
   Gérard Vallée
   1981 / xii + 114 pp.

THE STUDY OF RELIGION IN CANADA / SCIENCES RELIGIEUSES AU CANADA

1. **RELIGIOUS STUDIES IN ALBERTA**
   A State-of-the-Art Review
   Ronald W. Neufeldt
   1983 / xiv + 145 pp.

COMPARATIVE ETHICS SERIES/ COLLECTION D'ETHIQUE COMPAREE

1. **MUSLIM ETHICS AND MODERNITY**
   A Comparative Study of the Ethical Thought of Sayyid Ahmad Khan and Mawlana Mawdudi
   Sheila McDonough
   1984 / x + 130 pp.

Also published / Avons aussi publié

RELIGION AND CULTURE IN CANADA / RELIGION ET CULTURE AU CANADA
Edited by / Sous la direction de Peter Slater
1977 / viii + 568 pp. / OUT OF PRINT

Available from / en vente chez:
Wilfrid Laurier University Press
Wilfrid Laurier University
Waterloo, Ontario, Canada N2L 3C5

Published for the
Canadian Corporation for Studies in Religion/
Corporation Canadienne des Sciences Religieuses
by Wilfrid Laurier University Press